TRANSACTIONS

OF THE

AMERICAN PHILOSOPHICAL SOCIETY

HELD AT PHILADELPHIA

FOR PROMOTING USEFUL KNOWLEDGE

NEW SERIES—VOLUME XXVIII, PART I

APRIL, 1936

The Mammalian Fauna of the White River Oligocene—Part I. Insectivora and Carnivora

WILLIAM BERRYMAN SCOTT AND GLENN LOWELL JEPSEN

PHILADELPHIA:

THE AMERICAN PHILOSOPHICAL SOCIETY

104 SOUTH FIFTH STREET

1936

LANCASTER PRESS, INC., LANCASTER, PA.

THE MAMMALIAN FAUNA OF THE WHITE RIVER OLIGOCENE
PART I—INSECTIVORA AND CARNIVORA

By WILLIAM BERRYMAN SCOTT AND GLENN LOWELL JEPSEN
Princeton University

THE WHITE RIVER FORMATION

The name given these beds is taken from the White (or White Earth) River of Nebraska and South Dakota, along which some of the most striking exposures are to be seen. The strata cover a very large area in those states and continue, with interruptions, westward to central Wyoming and southward over northeastern Colorado. Outlying areas in North Dakota indicate probable former extensions of the formation. White River fossils have been reported in Oklahoma and Texas and also in Saskatchewan, Canada, but the beds in the southern U.S. cannot be identified. The strata, which are of no great thickness, seldom exceeding 200 to 300 feet, are very regularly stratified in nearly horizontal and undisturbed layers of fine clay-like sediments, transected at various levels and in different places by bands of cross-bedded sandstones and conglomerates. These bands, which may often be traced for many miles, have a lenticular cross-section and are rarely more than a few yards in width, often only a few feet, and constitute the "channel sandstones."

A varying quantity of volcanic ash is mingled with the clays, and some strata near the top of the series are made up entirely of ash, which is very fine and, apparently, was carried for long distances by the northwesterly winds.

The White River formation was long believed to be made up of a series of lake deposits, and the ancient lakes in which these sediments were laid down were repeatedly mapped. The lacustrine theory was first challenged by the late Dr. W. D. Matthew, of the American Museum in New York, who suggested that the beds were deposited by the wind, like the great areas of loess in northern China and the Pampas of Argentina. However, the extremely regular stratification over thousands of square miles seemed incompatible with the theory of aeolian deposits.

The aeolian theory has been generally abandoned in favour of the fluviatile theory, according to which the beds were laid down on the flood-plains of rivers and in their channels. It is clear that in the Oligocene epoch, the Great Plains over what are now very large areas of the Dakotas, Nebraska, Wyoming, and Colorado, were a low-lying, featureless country, traversed by a complicated net-work of streams, which were separated by very low divides. In the flood season, when all the streams overflowed their banks, the country was converted into a shallow temporary lake, which deposited fine mud and silt, and into which volcanic ash was showered from time to time. The channels were gradually filled up with sand and gravel, cross-bedded, as such deposits always are, thus forcing the streams to seek new courses, while repeated floods covered up the older channel deposits with the fine clays.

1

The fluviatile theory is not free from difficulties, but it offers the best solution of the problem that has yet been proposed, and it is strongly supported by certain unquestioned facts. The analogy of the Nile, and the manner in which it annually converts its valley into a temporary lake, is an excellent illustration of what a single river, without tributaries, can do in covering its whole flood-plain with the deposits of mud which have kept Egypt proverbially fertile for the last 6000 years.

The fossil bones themselves often show clear evidence that they lay on the ground,

Fig. 1.   Big Bad Lands of South Dakota.

exposed to the air, after the death of the animals, for they frequently show the tooth-marks of rodents, which could not have gnawed them when under water. Fossilized cases of insect larvæ frequently occur around the bones and these afford very strong proof of exposure to the air. In time of low water, when the flood-plains were dry, carcasses and skeletons accumulated, especially around water-holes, and were more or less mauled and dragged about by carrion-feeders, until the separate bones were scattered and gnawed by small rodents, or shattered by the powerful jaws of Carnivora.   When next the high-water stage covered the flood-plains, the newly dead carcasses, the more or less entire skeletons, the scattered bones, some complete, others broken up, were indiscriminately buried in the renewed deposits of silt and ash.

The channel sandstones are often richly fossiliferous, and they generally contain differ-ent animals from those which occur in the finer sediments, no doubt, because of a difference in habitat.   Remains of the presumably aquatic rhinoceros, *Metamynodon*, are generally

FIG. 2.   Bone of *Titanotherium*, gnawed by rodents before fossilization.   Princeton Univ. Mus.

found in the channels at one level, as are those of the strange ruminant, *Protoceras*, at another and higher level.

Many years ago the late Dr. J. L. Wortman proposed the subdivision of the White River formation into three parts, each characterized by a particular genus of mammals. Subsequently, Mr. N. L. Darton, of the U. S. Geological Survey, in accordance with the rule of the Survey, proposed geographical names for a twofold subdivision.   The relations are as follows:

WORTMAN	DARTON
Protoceras Beds ⎫	
Oreodon Beds ⎭	Brulé
Titanotherium Beds	Chadron

Both systems are in current use, though, as the *Protoceras* Beds are a channel deposit, it is necessary to designate the contemporary strata of ash and silt as the *Leptauchenia* Beds. Darton's division is made on lithological and structural features and, for palaeontological reasons, it is useful to recognize the Upper and Lower Brulé.

The Chadron substage (or *Titanotherium* Beds) is separated from the Lower Brulé (*Oreodon* Beds) by an erosional unconformity, which is not everywhere recognizable; the substage, in turn, is subdivided into three zones, the separate faunas of which still remain to be identified, except locally.

The White River Bad Lands were, according to Parkman, discovered in 1746 by the brothers La Verendrye, French-Canadian explorers, who had made their way westward as far as the Bighorn Mountains in what is now Wyoming, and their successors among the Canadian *Voyageurs* called these singular areas *"Mauvaises Terres à traverser,"* adopting a Sioux term. The abbreviated English translation, "Bad Lands," is in familiar use all over the arid and semi-arid West. The term refers to the extraordinary topography, which has been carved from horizontal strata of soft rocks by the action of the weather, especially rain, frost, and wind. Dr. F. V. Hayden, one of the first geologists to visit this region, wrote of it: "The beds of the White River present some of the most wonderfully unique scenery in the world. Over an area of about 100 miles from east to west, and 50 to 60 from north to south, they have been so worn and cut by streams, rains, and other atmospheric agencies, that they form one continued series of gullies, or dry gorges, with here and there isolated peaks and columns, looking much like steeples or towers, giving to the whole the appearance of the ruins of some ancient city." (Hayden in Leidy '69.)

The area which Dr. Hayden thus described is that generally known as the "Big Bad Lands," but the five or six thousand square miles so designated are far from representing the whole extent of the White River beds. In addition to separate, outlying areas, such as those in North Dakota and Saskatchewan, the White River formation extends over an immense area, where it is for the most part concealed under grasslands, or buried beneath Miocene, Pliocene, and even Pleistocene deposits; the actual area occupied by these beds has never been computed, but must be very great.

For almost a century the Bad Lands of the White River have been famous for the many and beautifully preserved fossil remains of mammals which have been there collected. The first printed account was published in 1846 by Dr. Hiram Prout of St. Louis. Soon after that, in 1847, Dr. Joseph Leidy described certain fossils which had been collected and sent in by employees of the American Fur Company. There followed a succession of collectors, most of whom gathered fossils incidentally, having other purposes in view. Dr. David Dale Owen and his assistant, Dr. John Evans, in 1849, Mr. T. A. Culbertson in 1850, and especially Dr. F. V. Hayden, who made several collecting trips to the Bad Lands, were the pioneers in scientific collecting. All the White River fossils were found lying on the ground and were obtained "without further exploration than picking them up from the surface of the country," in Dr. Leidy's words. Such material was, of course, more or less fragmentary.

Elsewhere, Leidy remarked: "The Mauvaises Terres collections of fossils rarely contained any considerable portions of skeletons preserved in continuity in masses of matrix. . . . They have been reported to exist and in future, when in the progress of opening up the country greater facilities will be afforded, they will no doubt be obtained and brought to the investigation of the student," a prediction which has been most amply and richly fulfilled.

For nearly a generation, all the White River fossils which were obtained were entrusted, as a matter of course, to the most competent hands of Dr. Leidy, an extremely important state of affairs for the development of American palaeontology. A succession of short papers in the Proceedings of the Academy of Sciences of Philadelphia, was followed by two magnificent quartos, the first published by the Smithsonian Institution in 1853, and the other by the Philadelphia Academy in 1869. This second and much larger volume, entitled *The*

Extinct Mammalian Fauna of Dakota and Nebraska, is the solid foundation upon which all subsequent work dealing with the White River fauna has been erected.

A new era in the history of the investigation of the White River, its stratigraphy and its fossils, began when, in 1873, Professors Cope and Marsh entered the field, both literally and metaphorically, for they made several collecting expeditions in person. The assistants whom they trained developed the collecting of fossil vertebrates into a fine art. Dr. J. L. Wortman, who began his career with Cope, and Messrs. J. B. Hatcher and O. A. Peterson, who first worked with Marsh, completely revolutionized the old methods, immensely increasing the quantity, and, what was even more important, the completeness of the material obtained. Instead of weathered skulls and fragmentary limb-bones, they brought in entire skeletons in astonishing numbers, so that it is now possible to supplement and extend Leidy's work in the most gratifying way, a way, indeed, which as noted above, was exactly predicted by him.

In Leidy's day and for some time after that, it was customary to refer the White River formation to the Miocene epoch of Lyell's original classification, but Cope, whose student period in Paris had made him familiar with the Tertiary geology of France, insisted that the White River fauna was much more nearly equivalent to the French Oligocene. As the latter term was seldom used in this country, Cope's opinion made slow progress, but its obvious correctness and consistency finally prevailed and the Oligocene reference is now in general use.

In the typical South Dakota area, the White River beds lie unconformably upon the eroded and worn surface of upper Cretaceous strata. Much the largest area of the Cretaceous involved is the marine Fort Pierre (or Pierre Shales) but there is also contact with the uppermost division of the continental Cretaceous, the Lance stage. Farther west, in Wyoming, the White River rests, also unconformably, upon the upper Eocene (Uinta stage) the lowest Oligocene of the Duchesne River stage being absent there.

The senior author of this report has long cherished the ambition to prepare a monograph of the White River mammalian fauna and make use of the immense collections stored in the various museums to fill out and complete the admirable sketch which Dr. Leidy gave to the world in 1869. It was, therefore, with peculiar pleasure that he received from The American Philosophical Society a grant which made it possible to visit and study the principal collections of White River fossils, in company with Mr. R. Bruce Horsfall, whose work in preparing the plates and text illustrations needs no eulogy. Accordingly, beginning in May, 1934, visits were made to the National Museum in Washington, the American Museum of Natural History, New York, the Carnegie Museum, Pittsburgh; in Chicago the Field Museum of Natural History and the Walker Museum of the University of Chicago; the State University at Lincoln, Nebraska, the Colorado Museum of Natural History at Denver, and the Museum of the State School of Mines at Rapid City, South Dakota. In October of the same year the pilgrimage was resumed, this time to New England, where the museums of Amherst College, Yale and Harvard Universities were visited and, finally, that of the Academy of Natural Sciences in Philadelphia.

At all of these institutions the visitors were received with the utmost kindness and cordiality and the treasures of each collection placed unreservedly at their disposal. The list of those to whom the authors are particularly indebted for assistance of every sort and, more especially, for the great number of fine photographs which were most generously contributed toward the preparation of this work, is a long one. Sincerest thanks are due to Professors R. S. Lull, Director, and Malcolm Thorpe, of the Peabody Museum, Yale University; to Professor F. B. Loomis, of Amherst College; to Professor A. S. Romer and Director Thomas Barbour, of the Museum of Comparative Zoology, Harvard University; to Professor H. F. Osborn, and Dr. Walter Granger, of the American Museum of Natural History, New York; to Director A. Avinoff and Messrs. Kay and Burke, of the Carnegie Museum, Pittsburgh; to Director Simms, Dr. Nichols and Messrs. E. S. Riggs and B. Patterson, of the Field Museum, and to Mr. Paul Miller, of the Walker Museum, Chicago; to Professor E. L. Barbour, of the Morrill Museum, University of Nebraska; to Director J. D. Figgins, of the Colorado Museum, Denver; to the late President C. C. O'Harra, Mr. J. D. Bump and Dr. George Hernon, of the School of Mines, Rapid City and, finally, to Mr. C. W. Gilmore, of the National Museum, Washington.

As the magnitude of the work became clear, it was an obvious necessity to gain assistance and so the senior author's colleague and one-time student, Professor Glenn L. Jepsen was invited to collaborate. Dr. Jepsen's long familiarity with the White River formation and its fossils, which began in his boyhood in Rapid City, at the very gateway of the Big Bad Lands, has made his assistance particularly valuable.

Until his premature and lamented death, in March, 1935, Professor W. J. Sinclair co-operated in the most helpful and efficient manner in the preliminary work indispensable to the preparation of this monograph. Dr. Sinclair spent many seasons in the White River Bad Lands and the great collection from that formation in the museum of Princeton University is a monument to his skill as a collector and preparator. To him and to his predecessor, J. B. Hatcher, the Princeton collection of fossil mammals owes almost everything.

THE WHITE RIVER FAUNA

Except for the tortoises, the overwhelming majority of White River fossils are of mammals. There is an interesting assemblage of reptiles; an alligator, several lizards and a few tortoises occur. While the species of tortoises are few, individually they are most abundant. A few skulls and bones of birds have also been identified, and many well-preserved eggs have been found, but with these lower vertebrates we are not here concerned.

The mammals were present in great numbers and diversity, as is well shown by the comparative table of families and genera which now inhabit all of North America north of the Neotropical Region, on the one hand, and those which have been found in the White River, on the other. No one can suppose that the latter list approximates completeness, or that many genera will not eventually be added to the list, to say nothing of many more, which altogether escaped fossilization. Even this undoubtedly incomplete list speaks eloquently of the great richness and diversity of the White River mammalian fauna; that the number of families should be so much in excess of those now living in North America, outside of the Neotropical part, and the number of genera so nearly the same, is surprising in view of the great discrepancy in the areas compared. On the one hand, there are a few thousand square miles of the Great Plains region, and, on the other, the millions of square miles of the whole vast continent. One factor makes any exact comparison impossible and that is the omission of the Cheiroptera from the table, because of the failure to find any fossil bats in the White River beds.   There can be no reasonable doubt that bats were represented in Oligocene North America, as they were in Europe, even in the Eocene, but bats are very rare as fossils and their absence from any particular Tertiary formation is without significance.   Another element of uncertainty is the number of titanothere genera which should be recognized, opinion and practice on this head varying within wide limits.

The determination of the *species* of fossil mammals is an exceedingly difficult problem, which is still far from solution, a regretable fact, which results from many causes, the principal one of which is lack of sufficient material.   Fossils give no information regarding growth of hair, size and shape of ears, colour, pattern and other external characteristics which are so largely relied upon in the discrimination of Recent species.   Rarely, too, can a sufficient number of skeletons be assembled to make plain the degree of individual variation within a given species.   The naturalist who deals with existing mammals requires a suite of specimens for the establishment of a new species, whereas many fossil species have been founded on a single tooth.   More than three hundred and fifty species of White River mammals have been named, but this number is far too large and will eventually be drastically reduced.   Existing animals teach that only a few species of a given genus can co-exist at the same time and place.   Each species, often each subspecies, has its own range and, except for overlapping borders, does not ordinarily share that range with another species, or subspecies.   For example, in G. L. Miller's *List of North American Recent Mammals*, the genus *Sigmodon*, or cotton rats, has 10 species, one of which, *S. hispidus*, has 23 subspecies. *Neotoma*, the wood-rats, has 27 species, but only two of these, *N. floridana* and *N. cinerea*, have as many as 10 subspecies each, the number of subspecies being thus disproportionate to the number of species.   Taking a few of the subspecies of *Sigmodon* as examples, their geographical ranges are given as follows: (1) North Carolina to Northern Florida and South Eastern Louisiana, (2) Eastern part of Florida peninsula, (3) extreme southern part of Florida, (4) Big Pine Key, (5) Eastern Texas and north to Kansas, (6) Rio Grande and Pecos Valleys, (7) Upper part of Gila Valley, Arizona, etc. . The range of the species is from Kansas to Panama, but each subspecies has its range to itself.

The element of time, with which the present geographical distribution of mammals is not concerned, plays a most important, if not always recognizable, part in the distribution of fossils.   It is not always feasible to say with certainty which species were literally contemporaneous and which were successive and such differences must be determined before a list of strictly contemporaneous species can be drawn up.   Five species of a genus, for instance, might be an impossible number to exist together in one area, while, if they rapidly succeeded one another in time, there would be no problem of co-existence.

In dealing with the palaeontology of mammals, species are comparatively unimportant; the significant unit is rather the genus, but that is merely because generic distinctions are so much more plain and easy to make.   Were the necessary information available, the species would be much the more important category.   In this monograph relatively little attention will be paid to the species, because of their generally indeterminate nature, and no attempt will be made to draw up a revised list of species under each genus, for such an attempt would be premature and of little value.   What can be said with confidence is that far too many species have been made, but, for the present, little more can be done than to list them and leave to the future the task of eliminating superfluous names.

In making the comparison between the Recent and the White River faunas, as to their relative abundance and diversity, there is one modifying factor, of which sight must not be lost, namely, the climatic one.   The climate of the Oligocene was very much milder and less extreme than that of the present and was much more uniform over vast areas of the continent.   The land-connection between northeastern Asia and northwestern America was on the site of Bering Sea and yet it served for back and forth migrations of southern mammals and brought about a remarkable similarity of mammals between Mongolia and North America in the Eocene and Oligocene.   That implies that mammals were then spread over the continent much more uniformly than now.   Hence a comparatively small area would give a more comprehensive picture of the life of the times than would be the case now. Even so, however, there must have been local differences and if we could know the Oligocene mammals of Greenland and Labrador, the list would be much longer.

In his introduction to Leidy's great monograph of 1869 Dr. Hayden made a prediction which, happily, has been completely falsified.   He wrote: "Many scientific men have said to me, in a few years these fossil remains will be exposed even more abundantly than now, by the erosive action of atmospheric agencies, so that we may look for novelties for a half-century to come. . . . The specimens are scattered over a large district, and are not abundant, and when once gathered from a certain area, the evidence seems to me clear that fifty years would do very little toward replacing them by the erosion of the beds in which the fossils are found.   No amount of excavation would be productive of important results" (p. 19).   No one would have been better pleased than Dr. Hayden to learn that his forecast was unduly pessimistic; nearly seventy years have passed since he made it, yet the "novelties" continue to come in.   How much the collectors have accomplished since 1868, the following comparative lists will make clear.

FAMILIES AND GENERA OF NEARCTIC LAND MAMMALS

Recent	*Present White River List*	*Leidy's List of 1869*
	MARSUPIALIA	
1 Family, 1 Genus	1 Family, 2 Genera	None
	INSECTIVORA	
2 Families, 11 Genera	7 Families, 14 Genera	1 Family, 2 Genera
	CARNIVORA	
7 Families, 21 Genera	3 Families, 17 Genera	2 Families, 3 Genera
	RODENTIA	
8 Families, 51 Genera	6 Families, 12 Genera	3 Families, 4 Genera
	LAGOMORPHA	
2 Families, 5 Genera	1 Family, 5 Genera	1 Family, 1 Genus
	ARTIODACTYLA	
4 Families, 12 Genera	8 Families, 26 Genera	3 Families, 4 Genera
	PERISSODACTYLA	
None	8 Families, 21 Genera	4 Families, 5 Genera
	EDENTATA	
1 Family, 1 Genus	?1 Family, ?1 Genus	None
25 Families, 103 Genera	35 Families, 98 Genera	14 Families, 19 Genera

Both in North and South America it is often necessary to infer that certain genera which appear suddenly and unheralded in a given area were migrants, not from some other continent, but from a different part of the same continent.

For example, the artiodactyl family of the Merycoidodontidae has never been found outside of North America and the successive evolutionary stages of most of its subdivisions and ramifications may be followed in the continental Tertiary formations of the West from the upper Eocene into the lower Pliocene. One series, or tribe, of the family made its first appearance in the upper White River in the genus *Leptauchenia* and continued into the middle Miocene. Though this genus is an unquestionable member of the Merycoidodontidae, nothing has been found in the older Oligocene from which it could have been derived and the most probable explanation is that the series arose in some other part of the continent and spread westward, eventually reaching the Pacific Coast.

The South American edentates, or Xenarthra, present a series of similar problems; the group, whether ordinal, or subordinal, is a very natural one and, despite its variety and diversity, there can be no question that its various members, armadillos, glyptodonts, ground-sloths, tree-sloths and ant-eaters, are all closely interrelated and were descended from a common ancestry. Furthermore, no representative of the Xenarthra has been found outside of the South American (or Neotropical) Region in strata older than the Pliocene, when the upheaval of the Isthmian sea-bottom united the two continents and enabled the southern edentates to spread into North America.

Pre-pliocene mammal-bearing formations in South America are almost entirely confined to the far southern part of the continent and, while many evolutionary series may be very satisfactorily traced from the older Eocene, this cannot be done for the edentates, the superfamilies of which appear unheralded at various levels and much the same statement applies to the monkeys and rodents. Assuming with von Ihering that Tertiary South America was divided into two provinces, Archiplata and Archiamazonas, between which communication was relatively difficult, it would explain the facts in the history of South American monkeys, rodents and edentates and we may conclude that the principal development of these orders took place in the tropical Archiamazonas, from which successive waves of migration extended into temperate Archiplata. Two of the xenarthrous groups, the tree-sloths and the ant-bears, never did reach the Patagonian region, so far as is known, and this was probably due to climatic causes.

A considerable part of the North American mammalian fauna was derived at various geological dates by immigration from the Old World. Some of the migrants, having become established here, underwent a long course of development and ultimately gave rise to genera and species which were very unlike their Asiatic, African or European ancestors. In tracing the history of the various families and genera, however, it is frequently insufficient to establish an ultimate Old World ancestry, it is often necessary to postulate an intracontinental as well as an intercontinental migration.

INSECTIVORA Gray

The members of this somewhat heterogeneous group are but loosely connected by primitive characteristics, for they are the most ancient and primitive of placental mammals. They seem to have been derived from the Jurassic Pantotheria and to have become distinctly established in the Cretaceous period. Because of their somewhat negative characteristics, it is often difficult to identify insectivores from fragmentary remains and to distinguish them from other primitive mammals. There is much reason to think that Huxley was right in his belief that in the Insectivora are to be sought the beginnings of all the orders of placental mammals. The relative importance of the order has steadily diminished since the Paleocene and in White River times they were much more numerous and diversified than they are at present in North America, seven families as compared with two. Furthermore, the Oligocene genera lack certain characteristics which many modern genera display and which are frequently regarded as primitive, but are more properly to be considered as degenerate. Thus, in several modern insectivores the zygomatic arch is incomplete and in others there are vacuities in the bony palate, as in marsupials. None of the White River genera exhibit such characteristics.

All existing Insectivora and all fossil forms which can be assuredly referred to the order are small animals; the largest modern species, *Solenodon cubanus*, has a length of 18 inches and the little "shrew-mice" (*Sorex*) are the smallest of all mammals.

"It is tempting to regard them (*i.e.*, the Insectivora) as a 'genetic group,' or a superorder like the Edentata, or Ungulata. But that would be to forget that it is a question of the most primitive Monodelphia, which, on the one side, just because of their great phylogenetic antiquity, display relations to other Recent mammalian phyla. On the other side, they are the terminal ramifications of Mesozoic mammals, which became specifically differentiated, or degenerate, in various directions and long ago attained the final phase of their development. They were preserved from extinction by their small size, their hidden manner of life and their adaptation to an insectivorous diet" (M. Weber II, 115).

Superfamily TENRECOIDEA Hay

Family 1. APTERNODONTIDAE Matthew

This family is as yet restricted to the Oligocene of North America, and finds its nearest relations in the West Indian *Solenodon* and the Madagascar *Tenrec*. The skull is, in certain respects, entirely unique among mammals.

Apternodus Matthew

(Pl. I)

Apternodus Matthew, Bull. Amer. Mus. Nat. Hist., XIX, p. 202 (1903); *ibid.*, XXVIII, p. 33 (1920).

The best-preserved skulls and mandibles of this genus are those in the American Museum of Natural History, the Harvard Museum of Comparative Zoology, and the Museum of the University of Wyoming. A very full description was published by the late Dr. W. D. Matthew on the basis of the Wyoming specimen, and an even more complete one by Dr. E. Schlaikjer of the Harvard skull. These may be summarized as follows: The dental formula is $i\frac{2}{2}$, $c\frac{1}{1}$, $p\frac{3}{3}$, $m\frac{3}{3}$. The median incisors are somewhat enlarged, the laterals with minute heel. The canines are two-rooted, stout and recurved. The first premolar, above and below, has been lost, the second is small and one-rooted; p$\underline{3}$ and $\underline{4}$ are three-rooted and triangular, with high central cone and basal cusps at the angles. The upper molars are similar, but wider transversely and the outer marginal cusps are higher than the inner one.

In the lower jaw, p$\overline{3}$ is two-rooted and has a high, conical, sharp-pointed cusp, with cingulum around the base; p$\overline{4}$ is molariform and the molars are triangular and tricuspid, with minute heel.

Skull. The skull in size and proportions resembles that of the Madagascar *Ericulus;* the face is rather short, the cranial region long and cylindrical with prominent sagittal and occipital crests. The orbits are ill-defined, as the postorbital processes of the frontals are very small and the zygomatic arches are wanting. The unique characteristic of the skull is a large flat, rectangular plate on the side of the cranium, formed by the mastoid, the exoccipital and the squamosal, continuous with the outer end of the post-glenoid process and bounded by heavy prominent margins which all lie in the same vertical plane. The dorsal border is on a level with the base of the sagittal crest; the ventral border is below the level of the mandibular condyle when in place; the posterior margin projects somewhat behind the occipital condyle and the anterior one rises from the post-glenoid process. "The occipital crests end at the posterior upper corner of this plate. The superior and posterior margins of the plate are evidently a development of the lambdoidal crests. . . . The inferior crest is evidently composed of the united paroccipital, mastoid and post-tympanic processes."

The occiput is rectangular, broad, low and concave and is composed of the exoccipitals and supraoccipital only; the *foramen magnum* is wide and low, and the occipital condyles are small and widely separated. The basioccipital is wide and short; the periotic rests externally and posteriorly against the lower border of the mastoid plate above described, and lies in a large pit, which marks the site of the auditory bulla, that was probably membranous; the alisphenoid is not involved in the bulla, and the pterygoids are not prominent. The postglenoid process is broad and heavy and has no foramen. The bony palate is without openings, or *fenestrae;* it is not produced behind the molar teeth and has a prominent posterior border.

The mandible is short and deep and has a wide, heavy condyle and a coronoid process which is of moderate height in *A. mediaevus*, very broad and high in *A. gregoryi.* The angular process of the jaw is prominent and incurved and the masseteric fossa is deep.

Aside from the unique squamosal-mastoid plates, this genus agrees well in the dentition and character of the skull with the Recent families, Tenrecidae and Solenodontidae.

Apternodus mediaevus Matthew

Apternodus mediaevus Matthew, *loc. cit.*

A separate description of the species was not given.

MEASUREMENTS

M$\overline{2}$–$\overline{3}$	3.9 mm.	M$\overline{3}$, longit.	2.0 mm.
M$\overline{2}$, longit.	2.0	M$\overline{3}$, transv.	1.8
M$\overline{2}$, transv.	2.0	M$\overline{3}$, height.	2.9
M$\overline{2}$, height.	3.7	Depth of jaw	3.5

Horizon: Chadron.

Apternodus gregoryi Schlaikjer

Apternodus gregoryi Schlaikjer, E. M., Bull. Mus. Comp. Zool. Harvard, LXXVI, p. 6.

This species, founded upon the very perfect skull, with mandible, which is figured in

Plate I, differs from the type of the preceding one in having a relatively larger and heavier lower jaw, a much larger coronoid process, a stouter and more inflected angle and a much heavier condyle. There are also a number of differences between this skull and the University of Wyoming specimen, which was described and figured by Matthew under the name of *A. mediaevus*, but these are characters which might well be individual or sexual variations. The fact, however, that *A. mediaevus* is from the Chadron and *A. gregoryi* from the Brulé is decidedly in favour of a specific distinction.

Horizon: Lower Brulé.

MEASUREMENTS (Schlaikjer)

Skull, length c to occ. condyle	38.5 mm.	M$\underline{3}$, greatest length	1.0 mm.
Skull, width over post-tymp. proc.	19.5	Mandible, length i$\overline{2}$ to condyle	26.5
Skull, greatest length of plate	15.8	Mandible, m$\overline{1}$–m$\overline{3}$ length	7.0
Skull, greatest height of plate	13.1	Mandible, height of coronoid	13.5
Skull, height of occiput	11.0	Mandible, depth at m$\overline{2}$	6.0
Cranium, length antorb. ridge to cond.	32.1	Mandible, width of condyle	8.0
Face, length antorb. ridge to i$\underline{3}$	11.7	Mandible, p$\overline{4}$, ant.-post. length	1.6
Tooth-row, length c to m$\underline{3}$ incl.	16.0	Mandible, p$\overline{4}$ height	2.6
M$\underline{2}$, greatest length	1.8		

Apternodus altitalonidus sp. nov., Clark (Ms.)

Distinguished by the large heel of m$\overline{3}$, Princeton Mus. No. 13,774.
Horizon: Chadron.

Family 2.   SOLENODONTIDAE Dobson

This family is, at present, restricted to the West Indies, species of *Solenodon* occurring in Cuba and Haiti. A genus from the Chadron substage is referred to the family, though with some doubt.

Micropternodus Matthew

(Pl. II, Fig. 1)

Micropternodus Matthew, Bull. Amer. Mus. Nat. Hist., XIX, p. 204 (1903).

The original description, which has not been added to in any significant way, is as follows: "Dentition $\overline{3.1.3.3}$. Molars somewhat like those of *Centetes* [*i.e.*, *Tenrec*] in composition, with high trigonid and small, low talonid. Trigonid very wide transversely, with protoconid considerably overtopping para- and metaconids. Talonid with sharp posterior margin and low median ridge. Molars and especially premolars short, high and recurved; p$\overline{4}$ submolariform, with small anterior and internal trigonid cusps and strong basal heel. P$\overline{3}$ much smaller and simpler, with small heel and no other accessory cusps. P$\overline{2}$ small and one-rooted, canine small, incisors small, subequal. No diastemata except a slight one behind p$\overline{2}$. Jaw rather deep in front. Second molar slightly larger than the first, third much smaller."

Micropternodus borealis Matthew

Micropternodus borealis Matthew, *loc. cit.*

The specific description consists only in measurements.

MEASUREMENTS

Lower jaw, length m3̄ to incis. alveoli....	12.4 mm.	Lower molar, m1 longit...............	1.8 mm.
Lower teeth, p3̄–m3̄.................	8.4	Lower molar, m1 transv..............	1.9
Lower molars, m1̄–3̄.................	6.4	Lower molar, m1 height..............	2.7

Clinodon gen. nov. Clark (Ms.)

(Pl. II)

The following description is contributed by Dr. John Clark from his unpublished manuscript entitled: *The Stratigraphy and Palaeontology of the Chadron Formation of the Big Badlands of South Dakota;* the genus and species, as well as the preceding *Apternodus altitalonidus* should be credited to him. As the description is somewhat shortened, it is not put in quotation marks, but is, nevertheless, substantially as he has written it.

Teeth preserved are the lower canine, third and fourth premolars and first molar; the type is Princeton Mus. No. 14,197. The teeth are closely set and strongly procumbent, with no diastemata, except possibly between p2̄ and 3̄. Canine small, laniary; p2̄ probably one-rooted. P3̄ with principal cusp pointed conical, ovoid in cross-section and a small, sharp posterior basal cusp, which touches the principal cusp of p4̄ and is outside of the anterior basal cusp of that tooth. P4̄ tricuspidate; the principal cusp is large, very broad transversely, much compressed antero-posteriorly and curving inward as it rises from the alveolar border, and with anterior and posterior basal cusps. The first molar has a high, antero-posteriorly compressed triangle of three cusps, of which the external one is much higher than the two internal ones. There is also a low, tricuspidate heel.

In composition of the crown, *Clinodon* is very like *Micropternodus*, but differs from that genus in having the teeth crowded and compressed, not spaced apart, and in being strongly procumbent and incurved rather than vertical. It is, furthermore, about one-third larger than the type of the genus last named.

Clinodon gracilis sp. nov. Clark (Ms.)

Diagnosis as for the genus.

Superfamily ERINACEOIDEA Dobson

Family 3. LEPTICTIDAE Gill

Members of this family are by far the commonest and most diversified of White River insectivores, individually much outnumbering all the other families combined. The group may be traced back into the Paleocene and seems to have come to an end in the White River, leaving no descendants.

Ictops Leidy

(Pls. III, IV)

Ictops Leidy, Proc. Acad. Nat. Sci. Phila., 1868, p. 316.
Nanohyus Leidy, *op. cit.*, 1869, p. 65.

This is very much the most varied and individually the most abundant genus of the family, but, as is true of other very small mammals, the fossils are, for the most part, very fragmentary. However, several beautifully preserved skulls, with mandibles attached, are in the museums of the South Dakota School of Mines, Rapid City, of the University of Nebraska and of Princeton University and these have all been studied and made use of in drawing the plates; the skeleton remains very incompletely known.

Dentition.   The dental formula is: $i\frac{2-3}{3}$, $c\frac{1}{1}$, $p\frac{4}{4}$, $m\frac{3}{3}$, and undoubtedly the normal number of upper incisors is two, but in one of the Rapid City specimens there are three of these teeth present on one side, two on the other.

A. *Upper Teeth.*   The minute upper *incisors* have single roots and chisel-like crowns, without cingulum, or basal cusps.   The *canine* is relatively large, especially in the fore-and-aft dimension; transversely, it is thin and compressed and though the crown is not long proportionately, it is a conspicuous little fang, implanted by a single, broad root.

The *premolars* increase in size from p1 to p3 and p4 is molariform; the external borders of the premolars and molars are all in the same antero-posterior line, the narrowing of the muzzle being gradual and without offset.   This arrangement of the cheek-teeth is one of the greatest differences between *Ictops* and *Leptictis*.   P1 is extremely small and has a compressed-conical crown, implanted by two roots; though perfect in form, this tiny tooth can have had little or no functional importance.   P2 is similar, but much larger; its external face is slightly asymmetrical, the sharp-pointed apex being a little in advance of the median vertical line, making the posterior border slightly concave.   P3 is larger than p2 and has a minute posterior basal cusp and a prominent internal cusp, which is made crescentic by the transverse ridges connecting it with the external cones.   P4 is almost like a molar, having two external cusps, of which, in unworn teeth, the anterior one is the higher; the crescentic inner cusp has a cingulum on the posterior side.

The molars are of the primitive, tritubercular pattern, with two external, conical cusps and a crescentic internal one, its arms forming transverse ridges, which may be interrupted to form conules; a prominent posterior cingulum gives the crown a nearly rectangular outline.   The first and second molars are of nearly the same size; m3 is much smaller and has lost the postero-external cusp, in some individuals at least.

B. *Lower Teeth.*   The three *incisors* are exceedingly small and styliform: the roots are large and cylindrical, with minute conical, enamel-capped crowns.   The lower *canine* is very much smaller than the upper one and shaped more like an incisor, but has a larger enamel-covered crown and is set off a little from i3.   However, these eight teeth, six incisors and two canines, look much alike and are all procumbent, becoming more erect from i1 to the canine.   The *premolars* are very much like those of the upper jaw, except p4; they have acute, compressed-conical crowns and increase in size posteriorly.   P1 is one-rooted, the others two-rooted, and p1 is shaped like the canine, but is broader, both crown and root; p2 is much larger and has an incipient posterior cingulum; p3 is still larger and has minute anterior and posterior basal cusps; p4 is partially molariform and has a posterior heel.

The *molars* diminish in size posteriorly; each is composed of two parts, the anterior "trigonid," of two high cusps, and the heel, or "talonid" of 3 much lower cusps.   This is an exceedingly primitive type of molar which has hardly changed at all from those which were prevalent in the Paleocene and Eocene epochs.

Several remarkably perfect skulls have been found, in addition to a great number in various degrees of incompleteness; it resembles closely that of *Leptictis*, as described and figured by Leidy, differing in only a few details of minor importance.   In size, the common species of the Brulé substage, *I. dakotensis*, almost exactly equals *L. haydeni*. The skull is long, with cranial and facial regions relatively elongate, the cranium, measured from the ant-orbital border to the occipital crest, slightly exceeding the face. The upper profile is a very regular curve, with highest point in the anterior parietal region, whence it slopes gently back to the occipital crest, and forward in a much longer and steeper incline to the end of the nasals. The muzzle tapers into a very slender rostrum, which gives the head a characteristically insectivorous appearance. Though the brain-cast shows that the cerebral hemispheres are small, having hardly any frontal portion and leaving the cerebellum entirely uncovered, the cranium is well-vaulted and has a pair of parallel temporal ridges, which are low and rounded, though very conspicuous; they extend from the occipital over the whole length of the parietals and half that of the frontals, dying away against the low convexities of the forehead produced by the frontal sinuses. The orbit is hardly demarcated at all from the temporal fossa, the jugal having no postorbital process and the frontal merely an indication of one, and in some skulls, even that is lacking.

The *occiput*, or inion, differs considerably in various individuals and these differences may, or may not be characteristic of some of the many species of the genus which have been named. In particular, two types may be distinguished, one in which the occiput is relatively high and narrow, and the other in which it is low and broad; in the former the temporal ridges are more nearly parallel and are closer together, while the median notch of the supraoccipital crest is deeper. In both types the occiput has a median keel bisecting a flat projection, with a depression on each side of it. The condyles are small and are widely separated by the very broad and low *foramen magnum;* the paroccipital processes can hardly be said to be present at all, as they are the merest vestiges. The mastoid appears as the edge of a thin plate between the exoccipital and the squamosal and its process is but little less reduced than the paroccipital. In all of the skulls at our disposal, the supraoccipital is indistinguishably united with the exoccipitals.

The *parietals* are united into a single large element, which forms much the greater part of the roof and walls of the brain-case, which is widest at the line of the glenoid cavities and tapers forward to the postorbital constriction. The *squamosal* is also large and forms nearly all of the side of the cranial cavity not covered by the parietal; its posterior border forms the ventral half of the occipital crest, which, though distinct, is not prominent; the posttympanic process is small, but much better developed than the paroccipital and mastoid processes, and the entrance to the ear is a broad, shallow groove between the large and heavy postglenoid and low posttympanic processes.

The *tympanic* bulla was, evidently, very loosely connected with the skull and has been lost in the great majority of crania, but in one individual (Princeton Mus. No. 11,420) it is preserved, though displaced from its original position and lies embedded in the matrix inside of the right *ramus mandibuli*. In size and shape, it is not unlike a grain of wheat, but is relatively somewhat broader. Leidy has described the tympanic of *Leptictis* as follows: "It is comparatively small and is in the form of a triangular scroll, descending by its base from the side of the basisphenoid and basioccipital, and curving outwardly to terminate in a free point. It reminds one of the process developed from the alisphenoid, enclosing the front of the tympanic cavity in the Opossum. No auditory canal extends from the bulla in the fossil, but an open archway communicates with its interior from between the postglenoid and postauditory processes, as in the Opossum" ('69, p. 34). In another individual, No. 10,539, the bulla is in place and agrees with that of *Leptictis*.

The *alisphenoid* is relatively large and articulates with the squamosal behind, the parietal above and the frontal in front.

The *zygomatic arch* is complete and not very slender in proportion to the size of the skull. The jugal is long and is extensively overlapped by the zygomatic process of the squamosal; its anterior end is furcate, to receive the very short zygomatic process of the maxillary; the dorsal prong of the fork extends up to form part of the anterior border of the orbit, but is far removed from the lachrymal. The facial portion of the *lachrymal* is a small triangle at the antero-superior point of the orbit; it has a low, blunt spine, which covers the foramen, just within the orbit.

The *frontals* are not coössified, and though large, form but little of the cranium, the posterior part being narrowed by the postorbital constriction. The temporal ridges continue over from the parietal upon the frontals and end in the low protuberances formed by the sinuses. Anteriorly, the frontals are deeply emarginated by the nasals and have long, acutely-pointed nasal processes.

The *nasals* are long and very narrow splint-like bones which articulate with the maxillaries and premaxillaries for the full length of those bones and project beyond the premaxillæ probably for the support of a prolonged, flexible snout such as most insectivores have.

The *premaxillaries* are extended antero-posteriorly on the face, but have no nasal or frontal processes, the suture with the maxilla being almost vertical; in accordance with the very narrow muzzle, the palatine processes are very small.

The *maxillaries* are elongate and vertically low; they are highest at the margin of the orbits and thence the height diminishes to the premaxillary sutures, where the whole muzzle, vertically and transversely, becomes characteristically slender; the suture with the frontals and nasals forms a continuous curve, with the convexity upward. The tapering of the maxillaries into the slender muzzle is gradual, without constriction or offset. The infraorbital foramen varies much in position, perhaps a species character. The palatine processes are long, narrow and moderately concave in the transverse and fore-and-aft directions. The *palatines* form a triangle, with notched apex. The bony palate, as in all other White River insectivores in which the structure is known, is entire and without vacuities. The border of the posterior nares is raised and "forms a double festoon, with a prominent, median palate spine" (Leidy). The *pterygoids* have distinct hamular processes and narrow fossae.

The *mandible* is long, slender and delicate; the horizontal ramus is very thin and its ventral border is convexly curved, vertically deepest beneath the molars, shallowing forward to the incisors, where hardly any bone is left. The two halves of the jaw remain separate throughout life, but are in contact in a long symphysis, which extends back to $p\overline{2}$. The coronoid process is inclined steeply backward, narrowing upward and ending in a blunt point; it rises high above the condyle and makes the sigmoid notch very deep. The condyle is transversely extended and has a low position, not much above the level of the teeth. The masseteric fossa is large and well-defined.

The Brain. Several brain-casts are available for study and they present some features of interest which were first pointed out by Bruce more than fifty years ago. The cerebral hemispheres are small, especially in length; in width they are well-rounded and capacious. Anteriorly, they leave the olfactory lobes exposed and posteriorly, not only fail to cover the cerebellum, but diverge so strongly that the *corpora quadrigemina* must have been exposed also. There is some indication of convolutions, very shallow, longitudinal grooves representing the lateral sulci, as may also be seen in the brain of the existing Hedgehog. Even more obscure is a groove that may be the supra-Sylvian fissure.

The lobes of the cerebellum are not clearly distinguished in the cast, but both median and lateral lobes may be made out; the vermis is very distinct, but not prominent.

Considerable parts of the skeleton have been obtained, which do not differ greatly from those of the hedgehogs, and as in most Insectivora, the lower part of the slender fibula is anchylosed with the tibia. The *manubrium sterni* (Pl. III, fig. 2) is of interest; it is broad and boat-like for most of its length, contracting posteriorly to a cylindrical form, to articulate with the meso-sternum. The dorsal surface is slightly concave and on the ventral side there is a prominent keel, as in *Rhynchocyon* of modern East Africa, but the horn-like processes, which in the existing genus, project between the facets for the clavicles and the first pair of ribs, are not present, though the facets themselves are very distinctly marked. Such a manubrium suggests that *Ictops* was, to a certain extent, a burrower, and *Rhynchocyon* is described as "a nocturnal animal, which lives in burrows" (M. Weber).

Species. No less than ten species of this genus have been named, a number that will assuredly be reduced when fuller and better preserved suites of specimens shall have been obtained. Much the greater number of these names have been assigned to fossils from the Chadron formation of Montana, in which many more small mammals have been found than in the equivalent beds of Dakota, Nebraska and Colorado. The type of the genus, *I. dakotensis* Leidy, was derived from the lower Brulé, or *Oreodon* Beds, and in that substage two species, at most, are indicated. One of the best characterized of the species is

Ictops acutidens Douglass

Ictops acutidens Douglass, Trans. Amer. Phil. Soc., N.S., XX, p. 245.

In his second paper on this group Douglass adopted the diagnosis made by Matthew. "Dimensions fifteen per cent less than any of the Leptictidae from the Oreodon Beds. First upper premolar one-rooted, two-rooted in *I. dakotensis* and *bullatus*. Supra-temporal crests widely separated anteriorly and convergent posteriorly, instead of close together and parallel, as in all the later species. Upper molars and p4 more constricted between the inner and outer cusps than in any described Leptictid; cusps somewhat higher . . . than in any later species" (Matthew, W. D., Bull. Amer. Mus. Nat. Hist., XIX, p. 205 (1903)).

Horizon: Chadron.

Ictops montanus Douglass

Ictops montanus Douglass, Mem. Carnegie Mus. Pittsb., II, No. V, p. 215 (1904).

"Skull quite broad and heavy behind, with fairly large brain case. In front of the orbits a gradual constriction from all sides to form the rather long muzzle: Zygomatic arches deep, broad as seen from the side. Postglenoid process long and inclining forward toward the lower end. Squamosal extending downward in a broad mastoid process [*sic*] outside of the external auditory meatus. Depression at anterior end of zygomatic arch rather deep. Anterior cusp of p4 small and low" (Douglass, E. *loc. cit.*).

Horizon: Chadron.

Ictops intermedius Douglass

Ictops intermedius Douglass, *ibid.*, p. 217.

"The skull is smaller than that of *Ictops montanus* and less robust. Though the skulls are similar throughout, yet there are differences, either slight or more pronounced, in nearly every part. The most noticeable are the following: In *I. intermedius* the zygomatic arches are much more slender; the postglenoid processes smaller and shorter; the mastoid process does not extend downward so far and is entirely different in form, the posterior portion of the skull is narrow and the *foramen magnum* smaller. The teeth are shorter antero-posteriorly, but fully as wide" (Douglass, E., *loc. cit.*).

Horizon: Chadron.

Ictops tenuis Douglass

Ictops tenuis Douglass, *ibid.*, p. 218.

"This is larger than either of the specimens previously described. The most noticeable differences, besides the greater size, are the greater width of the palate between the teeth and the much greater width at the postorbital constriction. The teeth are wider in proportion to the length than in *I. montanus*. The muzzle is slender in proportion to the size of the skull. The anterior portion of the zygomatic arch is proportionally more like that of *I. intermedius*. The skull is one fifth broader than in the latter between the orbits and one-third broader at the postorbital constriction. Though the mandible is nearly one-third longer, it is no deeper at the last molar" (Douglass, E., *loc. cit.*).

Horizon: Chadron.

Ictops major Douglass

Ictops major Douglass, *ibid.*, p. 220.

This is the largest species of the genus found in the Chadron. The lower dental series is 32 per cent (nearly one-third) larger than that of *I. intermedius* and 10 per cent larger than *I. tenuis*. The tail was long and heavy, but that was probably true of all the other species of the genus as well.

Horizon: Chadron.

Ictops thomsoni Matthew

Ictops thomsoni Matthew, Bull. Amer. Mus. Nat. Hist., XIX, p. 207 (1903).

"A species closely allied to *Ictops acutidens*, but distinguished by smaller size, more compressed teeth and other characters of less importance. The metacone on all the molars is decidedly smaller than the paracone: in *I. acutidens* they are nearly, and in other Leptictidæ quite, equal in size on m$\underline{1}$-$\underline{2}$. . . . All these distinctions are exaggerations of the differences between *I. acutidens* and the Leptictidæ of the Oreodon Beds" (Matthew, W. D., *loc. cit.*).

Horizon: Chadron.

Douglass gives the following dimensions of the species of *Ictops* from the Chadron substage.

Douglass concludes his survey of the Chadron species of *Ictops* thus: "From what I have seen of the Leptictidae of the Oreodon horizon [*i.e.*, lower Brulé] I think that in all these species from the Titanotherium beds of Montana [*i.e.*, Chadron], unless it be *I. major*, the teeth are proportionately smaller" (Douglass, E., *op. cit.*, p. 223).

	I. montanus	*I.* intermedius	*I.* tenuis	*I.* acutidens	*I.* major	*I.* thomsoni
	mm.	mm.	mm.	mm.	mm.	mm.
Skull, length back of p1.........	48.5	45.0				
Skull, width at p2..............	8.0	6.0	7.6			
Skull, width at post-orb. constriction.......................	13.0	12.0	16.0			
Occiput, width	20.0	18.0				
Occiput, width over condyles.....	15.0	13.0				
Palate, width at p2.............	5.5	6.0	6.5			
Skull, width at front of glen. surface.......................	25.0	24.0	33.0			
Molar-premolar series, length.....	23.0	22.0	28.0			
Premolar series, length..........	15.5	15.5	21.0			
Molar series, length............	7.5	6.5	7.0		10.0	
Molars and p3 and 4, length.....	14.0	12.0	14.0		18.0	9.3
P1, length....................	1.3	2.0				
P2, length....................	2.1		2.8			
P3, length....................	3.5	2.5	3.8	3.5	5.0	
P4, length....................	3.0	2.2	3.1	3.3	3.8	
P4, width....................	2.8	3.3	2.9	3.3	3.4	
M1, length....................	2.5	2.2	2.2	3.0·	3.6	2.7
M1, width....................	3.5	3.5	4.2	3.8	4.3	3.9
M2, length....................	2.2	2.2	2.2	2.2	3.5	
M2, width....................	3.8	4.0	4.0	4.2	5.0	
M3, length....................	2.0	2.0		1.5	2.0	
M3, width....................	3.0	3.0		3.0	4.0	3.6
Mandible, depth under m3.......		5.0	5.2	4.5	5.7	
Lower dental series, length.......		25.0	29.0		32.0	
Lower molar series, length.......		7.0		8.5	10.0	
M1, length....................		2.5		2.5	4.0	
M3, length....................		3.0	3.0	2.5	3.2	
Femur, length.................					45.0	

Ictops dakotensis Leidy

(Pl. III; Pl. IV, Fig. 2.)

Ictops dakotensis Leidy, Proc. Acad. Nat. Sci. Phila., 1868, p. 316.

Nanohyus porcinus Leidy, ibid., 1869, p. 65.

Dr. Leidy's type of this species is so fragmentary, that it may belong to either one of the two variants of *Ictops* skull which have been found in the Brulé formation. Matthew's selection of the larger and more robust of these for his *I. bullatus* makes it necessary to restrict the name *I. dakotensis* to the smaller and more slender type, so long, that is to say, as two species from the Brulé are admitted. Several finely preserved skulls, with mandible attached, are known of this species, those in the museums of the University of Nebraska, Princeton University and the South Dakota State School of Mines, being the most complete.

In these skulls the occiput is relatively high and narrow, the temporal ridges on the parietal are close together and almost perfectly parallel and the muzzle is conspicuously slender. A third skull is still smaller and more slender, but the difference may be due to the juvenile condition of the animal. The most perfect skull in the Princeton collection is No. 10,539; a second very good one, No. 11,420, lacks the anterior part of the muzzle, which is broken off in front of p2. Both of these agree closely in the characters before mentioned and also in the reduction of the last upper molar, of which the postero-external cusp is vestigial.

MEASUREMENTS

	No. 10,537	No. 11,420
Skull, median basal length	56.0 mm.	
Skull, length over all	64.0	
Skull, length p1 to occ. crest (on horizont. line)	52.0	
Skull, width at p2	12.0	11.0 mm.
Skull, width at post-orbital constriction	13.0	13.0
Skull, width at front of glenoid fossa	25.0	23.0
Occiput, width at base	24.0	24.0
Occiput, height fr. basi-occ.	17.0	18.0
Upper molar-premolar series, length	23.0	
Upper molar series, length	9.0	9.0
Upper premolar series, length	13.0	
P1, ant.-post. diam.	2.5	
P2, ant.-post. diam.	4.0	4.0
P3, ant.-post. diam.	4.0	4.0
P4, ant.-post. diam.	3.0	4.0
Canine ant.-post. diam.	3.0	
Lower molar-premolar series, length	26.0	
Lower molar series, length	9.0	10.0
Lower premolar series, length	16.0	
P1̄, ant.-post. diam.	2.0	
P2̄, ant.-post. diam.	4.0	
P3̄, ant.-post. diam.	4.0	
P4̄, ant.-post. diam.	4.5	4.5
Mandible, extreme length from angle	49.0	
Mandible, height of condyle	15.0	
Mandible, depth at m3̄	7.0	7.0
Mandible, depth at c̄	3.0	

Horizon: Brulé.

Ictops bullatus Matthew

Ictops bullatus Matthew, Bull. Amer. Mus. Nat. Hist., XII, p. 55 (1899).

Dr. Matthew bestowed this specific name under the misapprehension that the other species of the genus were without ossified tympanic bullae. Such ossifications were probably present in the living animal in all the species of *Ictops*, as they certainly were in *I. dakotensis*, as also in *Leptictis*. Being very loosely attached to the skull, the tympanic bullae were generally lost in fossilization. Matthew's name may, however be retained for the larger and more robust species of the Brulé substage.

In this species the occiput is relatively wider and lower than in *I. dakotensis;* the temporal ridges are more widely separated than in the latter and are not straight and parallel, but pursue a slightly sinuous course. They are widest apart at the occipital crest, whence they converge forward very gradually to the postorbital constriction, where they are most closely approximated. From the constriction they again diverge forward and are quite widely separated anteriorly and die away upon the protuberances of the frontal sinuses. A very typical, though unfortunately incomplete, example of this species is No. 10,526 of the Princeton museum, which is in decided contrast to the two specimens of *I. dakotensis*, of which the description and measurements are given in the preceding section.

MEASUREMENTS

Skull, width at p3................... 17.0 mm.	M1, length........................... 4.0 mm.
Skull, width at post-orb. cons......... 14.0	M2, length........................... 4.0
Skull, width in front of glen. foss...... 32.0	M3, length........................... 2.0
Occiput, width at base.............. 27.5	P4, length........................... 3.5
Occiput, height fr. basi-occ........... 18.5	Mandible, depth at M3............... 8.0
Upper molar series, length........... 9.0	

Horizon: Brulé.

Leptictis Leidy

(Pl. IV)

Leptictis Leidy, Proc. Acad. Nat. Sci. Phila., 1868, p. 315.

The type-specimen upon which this genus is founded is a beautifully preserved skull, lacking only the mandible and, strange to say, it remains the only example of the genus that has been discovered, or, at least reported. This extreme rarity may possibly be due to the fact that *Leptictis* is a mutant, or merely an individual variant of *Ictops*, but this seems very unlikely and cannot be assumed without further evidence.

The diagnostic differences between the genera are two: (1) The third upper premolar, which in *Ictops* has a large, crescentic internal cusp and a small postero-external one, is in *Leptictis* a simple, compressed cone, without accessory cusps of any sort. (2) In *Ictops* the anterior premolars are in line with the *external* margin of the molars (including p4), the tapering of the muzzle being gradual and without offset, while in *Leptictis* the anterior premolars are in line with the *inner* side of p4 and the molars and the muzzle shows a distinct offset in front of p4. It is this feature which makes it very unlikely that *Leptictis* can be merely an individual variant of *Ictops*. In other respects the skulls and upper dentition of the two genera are identical.

Leptictis haydeni Leidy

(Pl. IV, Figs. 1–1c)

Leptictis haydeni Leidy, *loc. cit.*

Leidy gives no definition of the species other than the following measurements in lines, here converted into millimeters on the basis of 1 line = $\frac{25}{12}$ mm.

Skull, estimated basal length......... 60.4 mm.	Skull, least width of cran............. 14.6 mm.
Skull, length occ. crest to fronto-nas. sut.. 38.3	Skull, width at post-orb. proc......... 16.7
Skull, length lat. occ. crest to ant. side of c 60.4	Skull, width at ant-orb. margin....... 22.0
Skull, length inion to ant-orb. border... 38.3	Skull, width at infra-orb. foram....... 14.7
Inion, height....................... 17.7	Face, length fr. ant-orb. margin....... 30.0
Inion, width........................ 27.1	Face, width at m1.................... 18.7
Skull, breadth at zygomata........... 38.1	Face, width at canines............... 9.3
Skull, breadth at post-glen. proc....... 24.0	Cheek-teeth, length................. 23.0

Mesodectes Cope

Isacis Cope, Palaeontological Bulletin No. 16, p. 3 (1873).
Mesodectes Cope, Systematic Catalogue of Vertebrata of the Eocene of New Mexico, p. 30 (1875).

This genus is defined as being identical with *Ictops* except that the minute postero-external cusp of p$\underline{3}$ is lacking. It is very doubtful whether such a trivial character is more than an individual variation.

Mesodectes caniculus Cope

Isacis caniculus Cope, *loc. cit.*, 1873.
Mesodectes caniculus Cope, *loc cit.*, 1875.

Metacodon gen. nov. Clark (Ms.)

(Pl. II, Figs. 5, 5a)

The description is taken from the unpublished Ms. of Dr. John Clark on the Chadron Formation, previously cited. The alterations in the text which have been made are omissions for the sake of brevity.

"*Type:* Partial lower jaw, with p$\overline{4}$, m$\overline{1}$–$\overline{3}$, Princeton Museum No. 13,835. In general character much like *Leptacodon* Matthew and Granger, . . . one-third to one-fourth larger. Angle of jaw almost completely absent, condyle almost in line with lower border of ramus. P$\overline{4}$ submolariform; metaconid directly lingual to protoconid. . . . Molars similar to those of *Leptacodon;* protoconid is higher than metaconid, which is higher than paraconid. . . .

"The specimen is placed in a new genus because the extreme reduction of the angle and lowering of the condyle until it falls in line with the tooth-row may be reflections of skull differences more striking than would be indicated by the conservative Leptictid lower molars. It seems undesirable to extend the range of an insectivore genus from the Paleocene to the Oligocene on the basis of imperfect lower jaws. . . . All the evidence of the type would indicate that *Metacodon*, if it is a valid genus, is derived directly from *Leptacodon*."

Metacodon magnus sp. nov. Clark (Ms.)

No separate diagnosis of the species is given.

Family 4. ERINACEIDAE Fischer

At the present time this family is confined to the Eastern Hemisphere, occurring in all the continents of that hemisphere except Australia. The family (and even the subfamily Erinaceinae) was present in the Oligocene of North America and the Leptictidae, which can be traced back into the American Paleocene, might, without unduly straining taxonomy, be made a subfamily of the hedgehogs. Since White River times no member of the Erinaceoidea has been found in the Western Hemisphere.

Proterix Matthew

(Pl. II, Figs. 2, 2a)

Proterix Matthew, Bull. Amer. Mus. Nat. Hist., XIX, p. 228 (1903).

"Dentition $\underline{3}$–$\underline{1}$–$\underline{3}$–$\underline{3}$, i$\underline{1}$ enlarged, c$\underline{1}$ large, two-rooted, p$\underline{2}$ small, one-rooted; p$\underline{3}$ small, three-rooted, with well-developed deuterocone; p$\underline{4}$ large, molariform, with small hypocone.

M1 and m2 wider than long, quadrate, with two external and two internal cusps of about equal size and a small, separate, postero-intermediate cusp (metaconule), the antero-internal cusp (protocone) with a ridge running out towards the antero-external margin. M3 trihedral, small, not extended transversely, paracone and metacone equal and well-separated, no hypocone. Palate completely ossified, its posterior margin as in *Erinaceus*. Skull bones arranged much as in *Erinaceus*, a well defined sagittal crest; premaxillae not reaching frontal bones" (*loc. cit.*).

As Matthew remarks, *Proterix* is nearly allied to the ancestry common to the subfamilies, Erinaceinae and Gymnurinae. It is much more likely to have been an immigrant from Asia than to have been derived from types of the American Eocene. Indeed, nothing is known from the latter, which can be regarded, with any probability, as ancestral.

Proterix loomisi Matthew

Proterix loomisi Matthew, *ibid.*

The only specific definition is in the following

MEASUREMENTS

Maxillary dentition c–m3 incl.	18.4 mm.	M1 transv. diam.	4.8 mm.
Transverse width of palate incl. molars.	17.6	M2 ant.-post. diam.	2.9
Depth of skull, junct. post-orb. crests to		M2 transv. diam.	3.9
palate	16.8	M3 ant.-post. diam.	2.0
Molar series, length	7.9	M3 transv. diam.	2.8
M1 ant.-post. diam.	3.4		

Horizon: Brulé.

Superfamily SORICOIDEA Dobson

Family 5.   SORICIDAE Fischer

Shrews are rare as fossils in Tertiary formations and the most ancient American representative of the family is the White River *Protosorex*. The type of this genus was sent to the senior author of this monograph, for description, by the late Professor George Baur, of the University of Chicago, and since Professor Baur's death the specimen has been lost. Most unfortunately, no figure was given in the original description.

Protosorex Scott

Protosorex Scott, Proc. Acad. Nat. Sci. Phila., 1894, p. 446.

Maxillary dentition much as in *Sorex*, but with less reduced third molar and smaller internal cusps on last premolar. Mandible with *four* minute teeth between the molars and the large, procumbent incisors. In all other known Soricidae the number of such teeth is *two*, except in one species of *Myosorex*, which sometimes has *three*.

Protosorex crassus Scott

Protosorex crassus Scott, *ibid.*

This species is characterized by its relatively large size, the rather short and broad face, vaulted palate and straight alveolar border. The type specimen is of an individual past maturity, for all the facial sutures have disappeared. The upper surface of the fronto-nasal region is straighter, broader and more flattened than in *Sorex*. The zygomatic arches have already completely disappeared and the suborbital portion of the maxillary ends in a rounded surface. In advance of p$\underline{4}$, the muzzle is suddenly narrowed; the palate is vaulted and deeply concave transversely; it is broad between the molars, narrow in front of them; there are no vacuities. The posterior nares have the same shape and position as in *Sorex*, but differ in the raised and thickened front border. The anterior portion of the muzzle is slender and tubular and the narial opening is small. The horizontal ramus of the mandible is relatively stout and has a single conspicuous mental foramen beneath m$\overline{1}$. Condyle and angle are missing.

In the upper jaw the crowns of the anterior teeth are broken away, leaving only the roots. The first incisor was large and compressed like that of *Sorex;* this is followed by four minute, one-rooted teeth, the homologies of which are doubtful. The last premolar is as large as a molar, but much less complicated; it resembles that of *Sorex*, but the internal cusp is less expanded and basin-shaped. The molars have the same cuspidation as in *Sorex*, but the small m$\underline{3}$ is much less reduced.

The large lower incisor has lost most of its crown, but would seem to have been less completely procumbent than in *Sorex*. Following this tooth are four minute and closely crowded teeth, with compressed, chisel-like crowns, of which the first and the fourth are slightly larger than the others. The molars are like those of *Sorex*.

MEASUREMENTS

Upper dental series, length, excl. i$\underline{1}$.....	9.0 mm.	Face, breadth at orbits................	8.0 mm.
Upper molar series, length...........	4.0	Face, breadth at p$\underline{3}$..................	5.0
Lower dental series, length, excl. i$\overline{1}$.....	7.0	Mandible, depth at m$\overline{2}$...............	2.5
Palate, length......................	10.0	Palate, width at m$\underline{2}$..................	4.0

Horizon: probably Brulé.

Family 6.   TALPIDAE Fischer

Moles are better represented in the White River than are the shrews, no less than three genera having been referred to this family. With one exception, however, they are in a lamentably fragmentary state.

Proscalops Matthew

(Pl. II)

Proscalops Matthew, Memoirs Amer. Mus. Nat. Hist., I, p. 374 (1901).

"*Generic characters.* Premolars not less than three above and two below, all except the fourth very small and one-rooted. P$\underline{4}$ with strong crescentic deuterocone and rudimentary tritocone. Molars short-crowned, trigonal in outline, with crescentic paracone, metacone and protocone and very rudimentary hypocone on m$\underline{1}$–$\underline{2}$. Skull exceptionally short and broad at the back.

"Founded on a skull and jaws (A.M.N.H. No. 8949 *a*) from the Leptauchenia Beds. This is the first mole recorded from the American Tertiary. The shortness of the skull and width of the occipital region separate it from any of the modern genera. *Scalops* most nearly approaches it in form of skull, though not so extreme, but has quadrate molars and all the teeth much higher crowned. *Scapanus* and *Talpa* retain the trigonal molars, but differ in the details of their dentition and are much longer in the cranial region. *Condylura* has a very much longer skull, premolars two-rooted and spaced, occipital foramen narrowed transversely. The remaining genera of the Talpidae differ in the dentition. . . . From all of them it is most easily distinguished by the short, wide head with very small cranial cavity."

Proscalops miocaenus Matthew

(Pl. II, Figs. 4, 4*a*)

Proscalops miocaenus Matthew, *ibid.*

"*Specific characters.* Last upper and lower molar greatly reduced. P$\underline{2}$ minute, p$\underline{3}$ slightly larger, p$\underline{4}$ with rather strong crescentic deuterocone and very faint tritocone. Protoconid greater than metaconid on all the molars, paraconid internal."

"MEASUREMENTS

"P$\underline{2}$–m$\underline{3}$	9.5 mm.	Max. width of skull behind arches	15.4 mm.
M$\underline{1}$–$\underline{3}$	6.1	Length of skull, p$\underline{2}$ to condyles	20.0
M$\underline{3}$	1.1	Length of skull glen. fossa to cond.	5.1
Transv. diam. m$\underline{2}$	2.6	p$\underline{3}$ to m$\overline{3}$	9.1
Width palate at m$\underline{1}$–$\underline{2}$	9.4	m$\overline{1}$ to m$\overline{3}$	6.7
Width palate at p$\underline{2}$–$\underline{3}$	3.9	Depth of jaw below m$\overline{2}$	1.8"

Domnina Cope

Domnina Cope, Palaeontological Bulletin, No. 16, p. 1.

Miothen Cope, Synops. of new Vertebrata from the Tertiary of Colorado, p. 5, Washington, Oct., 1873.

This genus, with its two species, *D. gradata* and *D. crassigenis*, was believed by its describer to be a bat, but is now more generally called a mole. It was established upon such uncharacteristic fragments, as to be of no importance.

Geolabis Cope

Geolabis Cope, Rept. U. S. Geol. Surv. of the Territories, III, p. 807 (1884).

Established to receive a species represented by portions of two crania without molar teeth, which may be referable to *Domnina*. The dental formula appears to be i$\underline{3}$, c$\underline{0}$, p$\underline{3}$, m$\underline{3}$. The first upper incisor is larger than the others and strongly decurved; the first premolar follows a short diastema, is two-rooted and has a simple compressed crown.

"The absence of the canine tooth and the enlarged first incisor distinguish this genus from *Talpa* and constitute a resemblance to *Scalops*" (Cope, *loc. cit.*).

Geolabis rhynchaeus Cope

Geolabis rhynchaeus Cope, *ibid.*, p. 808.

"The muzzle of this species is long and narrow, with vertical sides and convex superior surface, which is nearly straight in profile. The depth is a little greater than the width, and the extremity is rather abruptly truncate downwards and forwards. The nasal orifice is exclusively anterior. The nasal bones are narrow, but widen regularly posteriorly. The palate is moderately concave. The premaxillary bone presents a rather wide external face.

The cranium expands just in front of the orbits, which are well defined in front and above by a rather flat, wide frontal region. . . .

"MEASUREMENTS

"Length, muzzle fr. front of orbit (No. 1) 10.0 mm.   Width, ext. face of prmx.............. 2.8 mm.
    Depth at middle................... 3.0   Width between ant. borders of orb. (No. 2) 7.0
    Width at middle................... 2.8   Inter-orbital width (No. 2)............. 6.0"

"The cranium of this species is rather smaller than that of the *Scalops aquaticus*" (Cope, *loc. cit.*).

Horizon: Not stated.

FAMILIA INCERTAE SEDIS

Family 7.   APATEMYIDAE Matthew

The members of this family, which form a clearly defined genetic series, extending from the Paleocene into the White River, are of altogether uncertain ordinal position.

Matthew's example of referring them to the Insectivora, is here followed for lack of any better solution of the problem, though it is probable that fuller knowledge of these exceptionally peculiar animals will require the erection of a new order, Apatotheria, for their reception. They have been variously referred to the insectivores, to the primates and to the rodents and in several family combinations, especially with the Plesiadapidae, to which there are many resemblances. How these resemblances are to be evaluated, whether they are due to relationship, or were independently acquired, remains to be determined, when more complete material shall have been collected. Few of the fossil Insectivora, whether real or nominal, are represented in the collections by anything better than fragments. The only specimen of *Sinclairella* consists of lower jaws and a complete but crushed skull.

Sinclairella Jepsen

(Pls. V & VI)

Sinclairella Jepsen, Proc. Amer. Phil. Soc., LXXIV, p. 291 (1934).

The dental formula is uncertain, not as to the number of teeth, but as to their homologies; the formula is more probably: $i\frac{2}{1}$, $c\frac{0}{0}$, $p\frac{2}{2}$, $m\frac{3}{3}$, but may be: $i\frac{2}{1}$, $c\frac{1}{0}$, $p\frac{1}{2}$, $m\frac{3}{3}$. The doubt, it will be observed, affects the upper canines and premolars.

Upper Teeth. The first *incisor* is very large proportionately, curved and with enamel on the anterior and posterior faces, reflected also upon the sides, which are mostly exposed dentine. On the anterior face the enamel extends into the alveolus to a point above the foremost premolar (p$\underline{3}$), while, on the posterior face the enamel band is much shorter and ends above the second incisor. The masticating surface is worn into a transverse groove on the soft dentine between the projecting enamel edges. The pulp-cavity is almost closed, showing that growth was limited.

The second incisor is represented only by the root, as the crown has been broken away on both sides. Whether the incisors present are the first and second, or second and third, of the original three, is not known.

The *premolars* appear to be two in number, but the foremost of the two teeth may possibly be the canine; it is here regarded as p3.   The tooth is implanted by two roots, which are strongly curved backward; the crown is high and sharp-pointed, oval at the base, concave on the inner side, convex on the outer; the anterior face is grooved and the enamel is thinner there than on the remainder of the crown; the posterior edge is sharp.   The grooved anterior face apparently received the blade-like part of the third lower premolar and, if so, formed a very efficient cutting device.

P4 also has two roots; the base is irregularly oval, wider behind than in front; there is a small cusp at the base of the posterior edge and a small postero-internal ledge.   The anterior face is grooved, but much less deeply than p3.

The *molars* are fundamentally tritubercular in composition, but m1 and m2 are made quadritubercular by the addition of a postero-internal cusp.   The first molar has a large, subconical principal internal cusp, the two external cusps are conical and of nearly equal size; a continuous, sharp cutting edge extends over the anterior and posterior faces of these cusps, and the external cingulum is broadest on the posterior half of the crown and has a number of small tubercles; there is also a short antero-internal cingulum; a minute conule appears between the anterior external and internal cusps.   At the antero-external angle of the crown is a small accessory cusp, or style, which gives the crown an asymmetrical shape different from that of the other molars.

M2 is shorter antero-posteriorly than m1, having the anterior style more external than anterior but wider transversely; the external cingulum is wide and its outer border forms an irregular ridge.   M3 is smaller than the other molars and retains the primitive, trigonal, tritubercular composition, with two external and an internal cusp in triangular arrangement; no postero-internal cusp is present and the low external cingulum occurs only on the anterior moiety of the crown.

Lower Teeth.   The single lower *incisor* is scalpriform and has a cutting edge, the anterior part of the tooth is curved but, for most of its length, it pursues a straight horizontal course. The crown is broad antero-posteriorly, compressed transversely and covered with enamel on the exterior side.   The root is closed, not growing from a permanent pulp.   The incisors, upper and lower, when in place, have a rodent-like appearance, but the upper one has a truncated edge, not scalpriform.

The foremost *premolar*, p3, is peculiar and characteristic throughout the family and underwent successive modifications.   The anterior part of the crown in the type specimen is broken away, but there is reason to think that it formed a high, thin blade and the preserved portion forms a horizontal ridge.   The single root, which is posterior in position, slopes downward and backward outside of the incisor.   From p3 to m1 the alveolar border of the mandible ascends steeply and, midway between these two teeth, has a small socket, from which the tooth (p4) must have been shed during the lifetime of the animal, for it is partially filled with bone.   The last lower premolar underwent successive reduction through the history of the family and, in the White River genus, had no doubt become vestigial.

The *molars* are of the tuberculo-sectorial type, but the anterior triangle, or "trigonid" is modified by the great reduction of the antero-internal cusp, which is a mere tubercle of enamel and would not be recognized as one of the primary cusps, were it not that in the successive genera of the family, from one geological stage to another, this element had been steadily reduced.   From the antero-external cusp a ridge passes down to the anterior edge of the crown, where it forms a small basal cusp.   The heel, which is not much lower than the trigonid, is a deep basin, enclosed by a ridge on which tubercle-like cusps are placed.   The molars increase in size from the first to the third, which has the heel greatly prolonged antero-posteriorly.

The *skull* is very remarkable for the great development of the cranial and corresponding reduction of the facial region.   No doubt, the apparent breadth of the brain-case has been increased by the vertical flattening which it has undergone, but making all due allowance for this, it is clear that no other White River mammal had a brain of comparable proportionate size.   It reminds one of the skull of *Tarsius* and other minute Primates and affords a strong argument against including the Apatemyidae among the Insectivora, in none of which has any such brain development been observed.

The temporal ridges, which are very conspicuous and resemble those seen in skulls of the Leptictidae, extend forward from the lambdoidal crest to the orbits, of which they form a raised border, and continue into the zygomatic arches.   This prolongation differs from the condition in the leptictids, in which the temporal ridges end against the protuberances made by the frontal sinuses.   The parietals, which seem to be coössified, are a very large element which forms nearly the whole of the roof and much of the sides of the cranial cavity, the frontals making up a very much smaller part.   In the posterior parietal region are several vascular foramina, six or more on each side, and others open upon the occipital surface above the condyles.   Even from its crushed state, it may be inferred that the occipital surface was originally low and wide.

The mastoid processes are uncommonly large, especially in transverse width; the tympanic is an incomplete ring, supported by a marginal wall from the basi-cranial bones. The glenoid fossa is large and unusually flat and must have permitted very free movement of the mandible in the fore-and-aft direction.   The zygomatic arch is complete and well developed and the jugal has a small postorbital process, which is incurved.

The frontals, which remain distinct, are much smaller than the parietals; the remarkably deep post-orbital constriction narrows the frontals, which have no postorbital processes; they expand again to form the broad, flat forehead, which is without protuberant sinuses and is slightly dished by the raised orbital borders.   Anteriorly, the frontals are deeply emarginated, to receive the peculiarly shaped nasals, but have no nasal processes, if the sutures in this region have been correctly traced.

The nasals are short and, for most of their length, they are very narrow and splint-like, but posteriorly, they expand to a breadth three or four times as great as that of their anterior part; each nasal is thus rudely hatchet-shaped, with the haft in front, the blade behind.

In correlation with the greatly reduced length of the entire facial region, the maxillaries are short, especially the preorbital portion, which is sharply constricted to form the narrow and short rostrum, to which the premaxillae also contribute.   The zygomatic process is prominent and the infraorbital foramen, which opens above m$\underline{1}$, is unusually large.   .

The premaxillaries are relatively large, their palatine processes form a considerable part of the hard palate and have long and narrow, but conspicuous incisive foramina.

The *mandible* is short, in correlation with the abbreviation of the whole facial region. The horizontal ramus is especially short, its antero-posterior diameter being less than that of the ascending ramus; its vertical depth is greatest at m$\overline{3}$ and is much less at the anterior end, where it is merely a sheath for the large incisor.   The posterior mental foramen is large and conspicuous and is placed beneath m$\overline{2}$; a groove leads forward from the foramen to the incisor alveolus.

The ascending ramus is broad in the antero-posterior direction; the coronoid is high, with steeply-sloping anterior border, which, in side view, conceals the hinder part of m3. The condyle is ovoid in shape, with its principal diameter dorso-ventral, and transversely it is narrow; it has a very inferior position, beneath the level of the molars. The angular process is short and sharp-pointed and is somewhat inflected. The masseteric fossa is large and deeply impressed, with prominent anterior and inferior borders. The two halves of the mandible are separate and the symphyseal area is rather smooth and without definite boundaries.

Sinclairella dakotensis Jepsen

Sinclairella dakotensis Jepsen, *loc. cit.*

As but a single species of this paradoxical genus is known, the only specific diagnosis which is practicable is that afforded by the dimensions.

MEASUREMENTS

Length of skull, over all	57.0 mm.	Alveolar rim to tip of I1	11.0 mm
I1–m3 approx. on alveo. border	25.8	I$\overline{1}$ to m3, length	27.0
I1 ant.-post. diam. at alveo.	6.1	Lower molar series, length	12.4
I1 transverse diam. at alveo.	2.8	M$\overline{1}$, ant.-post. diam.	3.5
M1 to m3, length	10.9	M$\overline{1}$, transv. diam.	2.4
Mandible, length over all	45.0	M$\overline{2}$, ant.-post. diam.	3.7
Mandible, depth below m$\overline{2}$	9.7	M$\overline{2}$, transv. diam.	2.8
I$\overline{1}$, length fr. tip to end of root	23.8	M$\overline{3}$, ant.-post. diam.	5.5
I$\overline{1}$, vert. diam. at alveo.	4.9	M$\overline{3}$, transv. diam.	2.6
I$\overline{1}$, transv. diam. at alveo.	2.4		

Horizon: Chadron.

The problem concerning the systematic position and relationships of the Apatemyidae has been fully discussed by the junior author in his paper: *A Revision of the American Apatemyidae and the Description of a New Genus*, Sinclairella, *from the White River Oligocene of South Dakota* (Proc. Amer. Phil. Soc., LXXIV, No. 4, 1934) and we are not in a position to add anything to the conclusions there stated.

The rapid evolution of the Apatemyidae during their known range, from the Paleocene Torrejon-equivalent of the Fort Union in Wyoming, to the Oligocene White River, is a striking example of early and great specialization. The discovery, in the White River strata, of a type of mammals which had been regarded as a distinctly Paleocene-Eocene group adds another element of interest to the fauna.

If, as several authors (among them Abel, Le Gros Clark, and Simpson) have conjectured, the Apatemyidae are more conveniently placed in the Primates than elsewhere, another order is added to the White River list, and the geological range of the Primates in North America will be extended into the Oligocene.

Order CARNIVORA Gray

The abundant herbivores of White River times, which were nearly all small, or of moderate size, supported a numerous group of beasts of prey, some of which were surviving types of very primitive character, while others were progressive and, though primitive and archaic from the modern point of view, had yet advanced much beyond their ancestors of the lower Oligocene and upper Eocene stages.

Suborder CREODONTA Cope

The creodonts are an extinct group of primitive flesh-eaters, which, throughout the Paleocene and Eocene epochs, were extremely abundant and diversified, especially in North America, continuing through the lower Oligocene, after which they were absent from this continent, though in Europe they persisted into the Miocene.   In the White River but a single family of creodonts, the Hyaenodontidae, occurs, of which the genera, *Hyaenodon* and *Hemipsalodon*, were immigrants from the Old World and the European *Pterodon* is also found in the lower Sespé of California, which is correlated with the lower Oligocene Duchesne River of Utah.   It is true that the family is extensively represented in the American Eocene, but the genera do not appear to have been ancestral to those of the Oligocene, which, as just stated, are much more nearly related to European and Mongolian forms.   In the White River, the hyaenodonts seemed to be holding their own in competition with the true Carnivora, or Fissipedia; at all events, they are equally abundant as fossils.   Eventually, however, they succumbed, whether to that competition, or to some other factor of extinction, cannot now be told.

The Creodonta either have no sectorial teeth, or else they have more than one pair of such teeth, which no fissiped ever has; the lumbar and posterior dorsal vertebrae articulate with one another by means of cylindrical, interlocking zygapophyses like those of artiodactyls; in the carpus (save in certain European species of *Hyaenodon*) the scaphoid, lunar and central bones are separate; the feet are, with rare exceptions, pentadactyl and the metapodials have a radiating arrangement like the sticks of a fan, and, with the exception of the Arctocyonidae, the ungual phalanges are cleft at the tip, as in most Insectivora. As Cope originally pointed out, the creodonts have many resemblances to the insectivores and he was for a time in doubt whether the two groups should be included in the same order.

Family HYAENODONTIDAE Leidy

The Oligocene representatives of this family form a small group of genera, mostly identical with those of Europe, which first appeared in America in the Duchesne and Sespé, when the true Old World *Hyaenodon* migrated from Asia.   The European *Pterodon* also occurs in the Sespé, but has not yet been found in the Duchesne River, no doubt because of the general scantiness of fossils in that formation.   In the Chadron substage of the White River, as displayed in the Swift Current area of the Canadian Province of Saskatchewan, occurs the very large *Hemipsalodon*, which is still but incompletely known and may, perhaps, be referable to *Pterodon*.   It has also been reported, but not yet described, from Nebraska.   The higher portion of the White River, upper Brulé, or *Protoceras-Leptauchenia* substage, has, as yet, yielded no representative of the family, nor has any been found in the John Day.   The long history of American creodonts, native and immigrant, came to an end in White River times.

Hyaenodon Laizer et Parieu

(Pls. VII, VIII, IX, XVI, XXII)

Hyaenodon Laizer et Parieu, Echo du Monde Savant, 1838, p. 254, subgenus of *Didelphis;*
    Ann. d. Sci. Natur., 2ᵉ sér., T. XI, p. 27 (1839).
Neohyaenodon Thorpe, Am. Journ. Sci., Ser. 5, III, p. 278 (1922).
    The skull of this genus is one of the commoner White River fossils and has long been

familiar; skeletons were brought to light much more recently and now the bony structure may be completely described.   Many species have been named, distinguished chiefly by size, and these are, for the most part, unsatisfactory because of the differences due to age, sex and individual variability, but some are demarcated by structural characters.

DENTITION

.   The dental formula is: $i\frac{3}{3}$, $c\frac{1}{1}$, $p\frac{4}{4}$, $m\frac{2}{3}$, complete, that is to say, except for the loss of the last upper molar; the unworn surfaces of the teeth have a coarsely wrinkled enamel that is characteristic of the larger species, in the smaller ones the wrinkling is fine or lacking.

Upper Teeth.   The *incisors* are simple, conical, sharp-pointed and slightly recurved; i3 is considerably larger than the others, but there are specific differences in this feature.

The *canines* are large, conspicuously long and sharp, with heavy, gibbous roots, much as in the bears; they must have been terrible weapons and were the sole means of offence, for the feet of these animals were relatively weak and the claws blunt.   In several respects the hyaenodonts were analogous to the wolves and the relative development of fang and claw was one of the resemblances.

The *premolars* are narrow, sharp-pointed, of compressed conical shape, and have not at all the massiveness of hyena premolars; the generic name therefore, was not felicitously chosen.   The first premolar is much the smallest of the series, but is not at all vestigial, for it must have been functionally useful.   The tooth is implanted by two roots and the crown, which has no accessory cusps, is asymmetrical in that the apex of the cone is in advance of the middle of the base, making the anterior edge short and straight, the posterior longer and more concave.   P2 has the highest and most sharply pointed crown of all the upper premolars and is much larger and more symmetrical than p1, but has no cingulum and only two roots.

The third premolar is decidedly more elongate antero-posteriorly than p2, but is not so high vertically; it has an internal cingulum and a third root, but no inner cusp.   A posterior basal cusp makes it imperfectly trenchant.

P4 is more completely sectorial and consists of two parts; the anterior cusp is high and thick, but sharp-pointed and the posterior trenchant ridge is very short.   Though the inner, third root is much more prominent than in any of the other cheek-teeth, there is no distinct internal cusp, merely a sloping buttress, which strongly suggests that it is the remnant of a large cusp.   The internal cingulum, likewise, is better developed than in any of the other teeth.

Both of the upper *molars* are carnassials; m2 is considerably larger than m1; both have lost the inner cusp which must have been present in the Eocene ancestors of the genus, retaining only the third root and the buttress which springs from it.   The anterior cusp of m1 shows that it was formed by the fusion of two cones, for in the little-worn tooth the two points are distinct and separate.   This is not seen in m2, in which the coalescence of the two cones is complete.   Both teeth have posterior cutting ridges, that in m2 considerably the larger.   M2 has the remarkable peculiarity of being implanted in the zygomatic process of the maxillary and is thus placed *outside* of the eyeball.

Lower Teeth.   The *incisors* are very small and closely crowded together, to make room for the swollen roots of the canines; in the unworn condition they are shallowly bifid.

The *canines* are like those of the upper jaw, but somewhat smaller. The *premolars* are much like those of the upper jaw, except p$\overline{1}$, which is smaller than any of the others, has almost lost its compressed-conical shape and is little more than an enamel-covered ridge, inserted by a single root. P$\overline{2}$ is larger, higher-crowned and much more distinctly conical; the tooth is perfectly simple, without an accessory basal cusp. P$\overline{3}$ is much larger and higher-crowned and has a distinct posterior heel, and p$\overline{4}$ is exactly like p$\overline{3}$, except for being considerably larger.

Of the three lower *molars*, the first is much the smallest and the third the largest; they are all carnassials, but m$\overline{1}$ is imperfectly so and shows a different mode of wear from the others. All these molars have completely lost the internal cusps, of which not a vestige remains, leaving only shearing blades. In m$\overline{1}$ the two cusps which form the trenchant blade are worn from the apices downward, leaving only a stump in old animals. M$\overline{1}$ has a heel which, though small, is larger than in the other molars. M$\overline{2}$ is an efficient sectorial blade of two thin, trenchant cusps, and a very small heel, which can have been of no use. The blade is worn on the outer side by the abrasion of m$\underline{1}$, which shears down past it. M$\overline{3}$ is very like the lower sectorial of the cats and consists of two thin, trenchant cusps, which are much larger than those of m$\overline{2}$ and there is no vestige of a heel. When the jaw is closed, the hinder cusp of m$\overline{3}$ projects upward behind the orbit, where there must have been a special pocket to receive it.

The teeth are very constant in form throughout the genus and, in the various species, differ in little except size.

Milk Dentition. Very young skulls are rare as fossils. The best available specimen is a juvenile skull, with mandible, in the Princeton Museum (No. 10,916) of *H. crucians*,

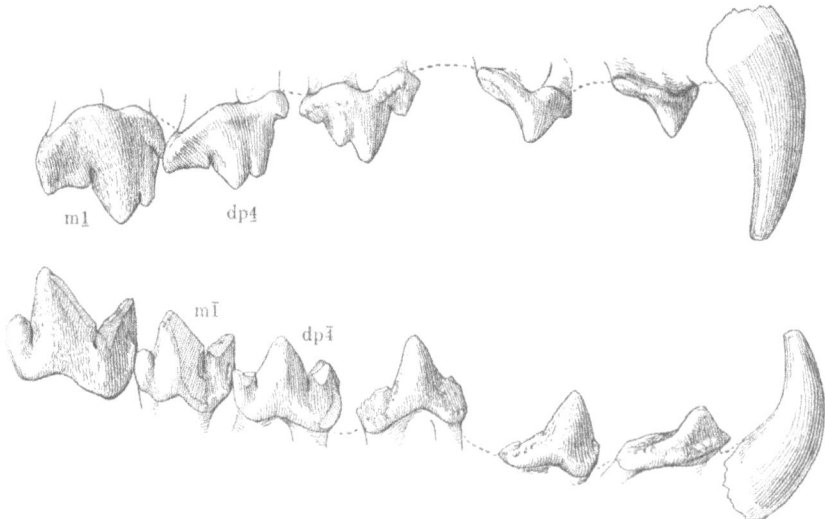

Fig. 3. *Hyaenodon crucians*, milk-dentition of the right side, $\times$ 4/3. dp$\underline{4}$, fourth upper milk-premolar; dp$\overline{4}$, fourth lower milk-premolar; m$\underline{1}$, first upper molar; m$\overline{1}$, first lower molar. Princeton Univ. Mus.

in which m$\underline{1}$, m$\overline{1}$ and $\overline{2}$ were already in use and show considerable abrasion, while in each jaw the last molar (m$\underline{2}$ and m$\overline{3}$) is visible, but not yet erupted.

Upper Milk Teeth. The *incisors* are arranged in a curved line, not in a nearly straight one, as are those of the adult; di$\underline{3}$ is much the largest of the series and is of simple curved form like that of a canine, di$\underline{1}$ and $\underline{2}$ are subequal and have faintly bifid crowns, with posterior heel. The *canines* differ from those of the second dentition merely in being rather smaller and more slender. The second canines are not erupted until all the permanent cheek-teeth are in place.

The first cheek-tooth, above and below, seems to be the first permanent premolar, freshly erupted and unworn; if so, these teeth have apparently no predecessors in the milk series; in that case the temporary formula will be: di$\frac{3}{3}$, dc$\frac{1}{1}$, dp$\frac{3}{3}$. This result was long ago reached by Osborn and Wortman (Bull. Amer. Mus. Nat. Hist., VI, p. 226 (1894)).

Dp$\underline{2}$ differs from p$\underline{2}$ chiefly in form, being relatively longer from before backward and much lower, so as to have a very different appearance; it also has an anterior cingulum, which p$\underline{2}$ lacks.

The third milk-premolar (dp$\underline{3}$) is altogether different from its permanent successor, having four external cusps instead of two, the additional ones being the anterior and posterior basal cusps, and the cingulum is somewhat more distinct. The tooth is implanted by three roots, but has only a vestige of the buttress, which is supported by the inner root.

Dp$\underline{4}$ is a sectorial and almost exactly like the first true molar; the anterior and median external cusps are connate, but they are of more nearly equal size and their apices are more distinct and wider apart. The posterior trenchant ridge is somewhat smaller and the whole crown decidedly lower vertically than that of m$\underline{1}$. There is no trace of an internal cusp, but the cingulum is well developed on the anterior face of the crown and faintly on the outer side of the posterior cutting blade. It is a tooth of this description, upon which Schlosser founded his *Pseudopterodon.*

Lower Milk Teeth. The *incisors* are even more closely crowded together than are the permanent ones and the median tooth of each ramus (di$\overline{2}$) is entirely behind and above the others, upon which it lies in procumbent attitude. As in the upper jaw, di$\overline{3}$ is larger than the others but does not exceed them so much; di$\overline{1}$ is the smallest of the series and has a simple, chisel-like crown. The crown of di$\overline{2}$ is relatively very long and styliform, cleft at the margin by a very shallow groove. As in the upper jaw, the canine is more slender than that of the second set. ˙

Dp$\overline{2}$ is present on one side of the mandible and p$\overline{2}$ on the other, thus allowing a direct comparison in the same individual; the two teeth are closely similar, but the temporary one has a somewhat lower and longer (antero-posteriorly) crown. The third milk-premolar is very like the fourth of the permanent series and consists of a high, pointed, compressed-conical cusp, with a low heel and minute anterior and posterior basal cusps. Dp$\overline{4}$ is a carnassial tooth, with cutting blade made up of two cusps, of which the anterior one is more conical and less compressed than in the sectorial molars; the heel, also trenchant, is relatively larger than in the latter. The sectorial pair is thus made up of the last milk-teeth above and below. -

THE SKULL

The skull is elongate; as in the Creodonta generally, the cranium is very long, but the face and jaws also are longer than is usual in the suborder. Seen from above, the skull has an hour-glass shape, the brain-case and forehead joining at the post-orbital constriction, which, in most species, is situated much farther behind the eyes than in such existing fissipeds as *Canis*, though the general appearance of the skull is rather wolf-like. The brain-case is relatively more spacious than in most creodonts, but much less so than in modern Fissipedia; it is pear-shaped, narrowing and shallowing forward, its roof sloping down steeply from the postorbital constriction to the occiput. In consequence of this condition, the posterior part of the sagittal crest and the occipital crest become very high and thin; the dorsal border of the sagittal crest is a remarkably straight, horizontal line, but the height of the crest above the brain-case diminishes rapidly forward and, at the postorbital constriction the crest is low, but distinct and divides into the temporal ridges, which are rugose and diverge abruptly into the post-orbital processes of the frontals. The preorbital vertical height of the face is remarkable.

The occiput has a characteristic shape; very broad at the base, it rapidly contracts and the dorsal half of it is very narrow, the crest forming a lanceolate arch, except in *H. crucians*.

In detail, it may be useful to make a comparison with the skull of the Coyote (*Canis latrans*) which, in general appearance and proportions, is not unlike that of *Hyaenodon* and *H. crucians* approximates it in size. Needless to say, there are many and striking differences in skull-structure between the Recent and the fossil animal, but some standard of comparison is needed.

The *basioccipital* is relatively shorter and wider than in *Canis*, not being encroached upon by the auditory bullae, which may not have been ossified. The part of the cranium which lies behind the line of the post-glenoid processes is very short, a primitive characteristic, especially in the large *H. horridus;* it is relatively longer in the small species. The basioccipital may, or may not, have a median keel; it is not constant in any species, but seems to be always absent in the smaller ones. The exoccipitals are broad at the base, narrowing much dorsally; the sutures with the supraoccipital are not visible in any of the available skulls.

As a whole, the *inion* is of very characteristic shape, very wide at the base, rapidly narrowing dorsally, to a degree that varies in the different species; in most of these the occiput narrows upward to form a pointed, lanceolate arch, but in *H. crucians* the narrowing is least, the dorsal part of the inion is relatively broad, with regularly rounded summit and, in consequence, the occipital crest, which is always prominent, is exaggeratedly so in this species. An instructive comparison is to view together from the occipital end, the skulls of *Hyaenodon*, especially *H. crucians*, and *Canis*. In the modern animal the well-rounded, spacious brain-case is broadly visible above and outside of the occiput; in the fossil the brain-case is completely hidden in rear view by the crest.

The occipital condyles are very peculiar in having the articular surface continued forward as a convexity almost at a right angle with the transverse part, the two together forming an L-shaped surface. The articular surface is also reflected over upon the medial side of each condyle, which must connect with the odontoid process of the axis. The paroccipital process is short and blunt-pointed; it is directed backward so nearly horizontally, that it must have been far removed from the auditory bulla. A narrow strip of the mastoid is exposed, wedged in, as it were, between the paroccipital and the posttympanic process of the squamosal, with which it becomes indistinguishably ankylosed; the mastoid process is very short, blunt and rugose.

The *basisphenoid* is very broad at its suture with the basioccipital, narrowing forward to its junction with the presphenoid, which is concealed, unless the palatines and pterygoids are removed. The relatively very large *parietals* seem to be ankylosed into a single element; at least no trace of the sagittal sutures is visible in any of the skulls examined; even in the very young animal, of which the milk-teeth were described on a foregoing page, the suture is not to be seen. The compound parietal forms nearly all of the cranial roof and much of the side-walls, the frontals adding but little, while the limits of the ali- and orbito-sphenoids are not clearly distinguishable in any of the specimens.

The *squamosal* forms a large proportion of the cranial wall below and behind the parietal, but extends little in front of the glenoid cavity. The postglenoid process has a concave anterior face, the distal end curving forward. The conjoined mastoid and post-tympanic processes are widely separated from the postglenoid, forming an arcade for the auditory meatus; an oval fossa, at the top of which the periotic may be seen, probably lodged some sort of an auditory bulla, but, if so, it was either not ossified, or so loosely attached as to be lost from all of the many skulls examined. The zygomatic process is surprisingly thin and weak, in comparison with the same bone in the modern fissipeds; it has a much shorter contact with the jugal than in *Canis*, tapering forward to a point.

The *jugal* also is slender and frail, though it broadens anteriorly and becomes fairly stout where it rests upon the maxillary; it has no indication of a post-orbital process. The zygomatic arch pursues an almost horizontal course, which is in marked contrast to the upward curvature seen in the fissipeds and the fragility of the bones is unintelligible in a beast of prey. The size and depth of the masseteric fossae on the mandible are proof of powerful masseter muscles and how such a frail-looking zygoma could withstand the pull of strong muscles is very puzzling. Most *Hyaenodon* skulls are found with the zygomatic arches wanting, but none is known which shows any evidence of arches fractured in the lifetime of the animal. The jugal extends to a contact with the lachrymal, these two bones forming all of the inferior and anterior boundary of the orbit.

The *lachrymal* is unusually large in its extension upon the face and within the orbit; each of these surfaces is many times as large as in *Canis*. The lachrymal foramen is concealed within the rim of the orbit.

Compared with those of *Canis*, or other modern fissiped, the *frontals* of *Hyaenodon* are small, for they cover very little, if any, of the cerebral fossa, the coronal suture usually coinciding with the postorbital constriction, whereas in *Canis*, for instance, this suture is far behind the constriction. The frontal projects much farther out over the orbit than in *Canis*, thus roofing it extensively and the prominent postorbital process is not the mere angulation that it is in the dogs. The orbits present laterally, without the forward slant which existing Fissipedia generally have. The forehead has a shallow median concavity, with large, low convexities on the sides, which cover what seem to be very extensive sinuses. The anterior emargination for the nasals is a wide-open notch, the form of which differs in the various species, but in none do the frontals have nasal processes.

The *nasals* are narrow and elongate, slightly convex longitudinally and strongly so transversely; they broaden posteriorly and are widest at the anterior end of the fronto-maxillary suture, whence they narrow to the hinder ends, which are points.   The anterior ends of the nasals are simplest in *H. mustelinus*, in which they form median, rather blunt points; the other species have these median points, which are shorter and blunter than in *H. mustelinus*, but also lateral processes along the premaxillary sutures, between which and the median points the bones are deeply notched.

The *maxillaries* are, in some respects, very peculiar; they contract anteriorly to form a slender muzzle, which is very wolf-like, sometimes more fox-like in appearance.   In front of the orbits the maxillaries have remarkable height, which is most obvious in the largest species, *H. horridus*.   The alveolar border is separated by a deep notch from the palate and thus arises the extraordinary feature of the last molar being inserted in the zygomatic arch, at least, in appearance, and thus adds to the paradoxical character of the arch, which in spite of its seeming frailty, was able to support all the stresses that were exerted upon it. The palatine processes are not unlike those of *Canis*, but the different form of the teeth, all of which are trenchant and none of which have inner cusps, gives the palate a different shape.   The infraorbital foramen is a narrow slit, placed above p3, but varying in relative distance from the orbit.

The *premaxillaries* are rather small, but stout and strong; the alveolar portion for the incisors is thick and on the side is depressed to form a groove for the reception of the lower canine, which is much more distinct than in *Canis;* the ascending ramus is short and has a much less extensive contact with maxillaries and nasals than in the modern genus.   The palatine processes are small, relatively most so in the smaller species, in which the incisive foramina are narrow slits; they are proportionately much wider in the larger forms.

Not the least peculiar parts of these very peculiar animals are the *palatines* and *pterygoids*.   The anterior part of the palatines, between the tooth-rows, is not unlike that seen in *Canis*, but the ends are broader and more abruptly truncate.   The extraordinary feature is the junction and coalescence of the two palatines and two pterygoids in the median line for their entire length, thus prolonging the nasal passage as a tubular canal almost to the line of the postglenoids and making the posterior nares open backward instead of ventrally. In some individuals the pterygoid plates of the alisphenoids unite in the middle line. Sometimes the posterior portions of the palatines are separated by a narrow slit, but this seems to be a matter of age and individual variation, not a species character, though it has been so regarded.

The only other existing mammal in which the posterior nares open so far behind their ordinary position is the Ant Bear (*Myrmecophaga*) and no one can imagine that *Hyaenodon* was an ant-eater.   What the significance of this very exceptional structure can be, is a subject for conjecture, since no direct evidence can be obtained and no similar animal is now living.   The unusual condition of the nasal passage has been compared to that seen in the crocodiles, though there is no other feature in the teeth or skeleton of the hyaenodonts to suggest amphibious habits.   Crocodiles and alligators are able to drown their prey by holding it under water, yet breathing freely themselves, so long as the tip of the snout is above water.   It does not seem likely that these predaceous mammals had the habits of crocodiles, yet, if not, no other explanation of the paradox is any more probable.

The *mandible* is very long and has a number of peculiarities; the two halves of the jaw ankylose early and, even in the youngest available skull, with milk-teeth (No. 10,916), the fusion is already complete and shows no trace of a suture. The symphysis is uncommonly long and extends back to p$\overline{3}$, or even to p$\overline{4}$; it is more or less concave on the dorsal side and antero-posteriorly strongly convex on the ventral.

The horizontal ramus is very long, but relatively rather slender and weak, in comparison with the size of the teeth, as appears most strikingly in the small *H. mustelinus* and is as inexplicable as are the slender and fragile-looking zygomatic arches. The jaw becomes very narrow at the symphysis and the lower canines are so approximated as to leave little room for the incisors. The ventral border of the ramus is slightly sinuous; from the angle to the forward end of m$\overline{3}$ the border is moderately concave, thence to the incisor alveoli, the curvature is gently and regularly convex. The outline is similar to that of the mandible in *Canis*, but in the latter, concavity and convexity are much more decided. The alveolar border is more concave than in the Coyote, m$\overline{3}$ rising high above the level which it has in the existing animal. Several small foramina appear on the anterior aspect of the symphysis and there are two or three conspicuous mental foramina on the side, the forward one beneath p$\overline{2}$ and the hindmost under p$\overline{4}$.   ·

The ascending ramus is very broad from before backward, but rather low vertically, though the coronoid process is well developed; the anterior edge of the process is much less vertical than in *Canis*, but slopes downward and forward, to merge in the alveolar process, for this edge is thin and has no *linea obliqua externa*, such as is seen in *Canis*. The coronoid is nearly erect and its posterior border curves downward gently into the condyle, without definite sigmoid notch. The proximal end of the process is somewhat thickened and rugose for the insertion of the temporal muscle. The masseteric fossa is a remarkably large and deep depression, especially in *H. horridus*, in which it extends forward beneath m$\overline{3}$ and has a raised crest bounding the antero-superior side. In the smaller species the fossa is relatively not so large. The condyle is greatly extended transversely, especially toward the inner side; the position of the condyle is remarkably low, well below the level of the teeth. The angular process is a short, broad hook which is rugose at the tip and slightly inflected, not enough, however, to suggest marsupial relationships. In length, it differs much in the various species; it is longest in *H. cruentus*, in which it projects well behind the condyle, shortest in *H. horridus* and *H. crucians*, in which it does not reach the vertical line of the condyle, and intermediate in length in *H. mustelinus*.

No part of the *hyoid arch* has been reported.

THE BRAIN

The brain of *Hyaenodon* is represented by several cranial casts and one of these, not a very favourable one, was long ago figured by the senior author (Scott, W. B., Journ. Acad. Nat. Sci. Phila. 2nd Ser., IX, Pl. 7, fig. 4, 1886). None of the casts is entirely complete, but by combining them, all parts of the brain may be made out. The brain, as a whole, and the cerebral hemispheres, in particular, are small. The olfactory lobes are large and are not covered over by the hemispheres at all, which also leave the cerebellum exposed, but conceal the corpora quadrigemina.

The cerebral hemispheres are roofed almost or quite entirely by the great parietal bone, the frontals hardly touching them, and thus, in the etymological sense, they have no frontal lobes. The cerebrum narrows forward, to a moderate extent and there it also has its least dorso-ventral diameter.   The temporo-sphenoidal lobe is very well developed and gives great vertical diameter to the posterior part of the brain, projecting down below the level of the optic chiasma, which is clearly displayed in one of the casts.   Sulci and convolutions are obscurely marked, but that is probably due to the space occupied by the brain membranes, which, in many mammals, prevent any close reproduction of the cerebral surface.   Indications of the Sylvian fissure may be seen and the supra-Sylvian and lateral sulci may be made out.   Between the two latter there is some sign of a third sulcus, but it is doubtful.   Apparently, this is a more highly developed brain than that of the White River dog, *Daphoenus*, or even of the lower Miocene *Cynodesmus*.   Perhaps, it was their superior intelligence that enabled these antique creodonts to maintain their footing so long against the more advanced and differentiated fissipeds of the White River, only to vanish before the end of the stage.

SPINAL COLUMN

Skeletons of *H. mustelinus* and *H. cruentus* are available for study and in the two skeletons the backbone is preserved, without a break, from the atlas to the end of the tail. Both of these have the same vertebral formula: C 7, D 14, L 6, S 3, Cd 26–7; the neck is rather short, body and tail long; the tail is especially elongate in *H. mustelinus*, in which, though the number of caudal vertebrae is the same as in *H. cruentus*, the individual vertebrae are relatively much longer.

The *atlas* is small, both antero-posteriorly and transversely; the inferior arch is a slender bar of bone, which is without the anterior tubercle that serves for the attachment of the *longus colli* muscle and is conspicuous in *Canis*.   The neural arch is considerably broader from before backward than the inferior one, and is smoothly convex, without trace of the neural spine.   The transverse processes are smaller in both dimensions than those of *Canis*, projecting but little behind the posterior cotyles, while in the modern genus they are produced considerably back of those surfaces.   In the latter also the atlanteo-diapophysial notch transmits the ventral branch of the first spinal nerve, but in *Hyaenodon* the notch is converted into a foramen, which opens into a pit common also to the oblique foramen.   The opening of the vertebrarterial canal, which in *Canis* is on the dorsal side of the transverse process, in the fossil pierces the root of the process and is invisible except when viewed directly from behind.

The *axis* has the shape common to nearly all land carnivores, but with some characteristic modifications; thus, from the ventral surface, near the hinder end of the centrum arises a prominent, plate-like hypapophysis.   The transverse processes are like flattened rods, broad dorso-ventrally, compressed and thin transversely and extending backward, very little outward, overlapping the processes of the third vertebra.   The pedicles of the neural arch are narrower, but higher than in *Canis* and the great neural spine, which is shaped like the old "broad axe," rises higher dorsally than in the Recent genus.

The *third* and *fourth cervical vertebrae* are almost exactly alike, with short, narrow centra, which bear hypapophyses upon their ventral surfaces; the transverse processes are much extended antero-posteriorly, but have no inferior lamellae.   The neural spine is short and broad on the third vertebra, longer and narrower on the fourth.   On the remaining cervicals, hypapophyses are lacking; the inferior lamella appears on the fifth and is still larger on the sixth, where it is of more regular quadrangular shape; from this the transverse process projects as a short, blunt rod, perforated by the canal for the vertebral artery. The seventh cervical, as is usual, has neither inferior lamella, nor arterial canal. The neural spines grow successively longer and narrower posteriorly, the seventh having the longest.

The *dorsal vertebrae* are not in any way peculiar; they have stout centra, which increase very gradually in fore-and-aft length posteriorly. The spines, in the larger species, are much as in the wolves, but in *H. mustelinus* they are more slender, somewhat as in the larger mustelines such as the Fisher (*Mustela pennantii*). The spines decrease in length posteriorly and incline backward to the eleventh or twelfth dorsal, which is the anticlinal vertebra and has a short, erect spine. Metapophyses, anapophyses, and the lumbar type of zygapophyses appear on the thirteenth vertebra.

The *lumbar vertebrae* are large and heavy, especially in *H. horridus*, and have short, broad, forwardly inclined neural spines, which increase in length posteriorly. The zyga-pophyses have the cylindrical, interlocking shape which is found in all known creodonts and which they share with artiodactyls and a few other mammals. Metapophyses, which appear on the last two dorsals, are prominent on the first three lumbars, become very low on the fourth and are lacking on the three succeeding vertebrae. Anapophyses, which are present on the last two or three dorsals, increase to a maximum on the second lumbar and then diminish, disappearing from the last two. The transverse processes which are sur-prisingly weak and slender, first appear on the second lumbar, increase in length to the sixth and shorten again on the seventh because the pelvis leaves insufficient space for them.

The *sacrum* consists of two or three vertebrae, of which only the first is in contact with the ilia. The neural spines are united only at their bases and are separate for most of their length; the transverse processes are fused into a continuous plate on each side.

The *caudal vertebrae* number 26 in the New York skeleton (*H. horridus*), 27 in the Princeton specimen (*H. mustelinus*), but the relative length of the tail is very different in the two species. In the larger one the tail is of moderate length and is about as in the Gray Wolf, but in the small animal it is proportionately very much longer and equals that of one of the large cats in proportionate length. The difference is not in the number, but in the length of the vertebrae. In *H. mustelinus*, the first four vertebrae have short centra and complete, well developed processes, transverse processes, zygapophyses, etc., from the fifth caudal backward, these are gradually reduced to mere tubercles and ridges. At the same time, the centra increase in length and diminish in diameter to the sixteenth, after which they are gradually reduced in all dimensions to the final rod at the tip.

The *ribs* are like those of the Carnivora generally. The first one is the shortest and broadest, the second and third are successively longer and narrower; in *H. mustelinus* all the ribs behind the third are very slender and rod-like, not strongly curved and forming a narrow thorax. In *H. horridus* the ribs are somewhat broader and more as in the wolves. Tubercles, articulating with the transverse processes, are well developed back to the eleventh pair of ribs, but the last three pairs are without them.

The *sternum* is not represented in the skeleton of *H. mustelinus*, but in the *H. horridus*, of the American Museum, the manubrium and six or seven segments of the mesosternum are preserved. In the Princeton skeleton of *H. cruentus* there are the manubrium and six additional segments. The manubrium is prolonged much in advance of the surfaces for the attachment of the first pair of ribs and its principal diameter is dorso-ventral, while, transversely, it is narrow and compressed. This type of manubrium is pinniped rather than fissiped in character.   The mesosternum must have consisted of at least seven segments, which are relatively rather short and heavy, four sided and slightly contracted in the middle and widening at the ends.   They are shorter and heavier than those of *Canis*.

LIMB GIRDLES

The *scapula* is known in three of the species and shows some specific differences; in the larger ones, *H. horridus* and *H. cruentus*, the shoulder-blade is relatively small.   There is hardly any coraco-scapular notch, the coracoid border curving gently into the straight dorsal moiety; the supra-scapular border is slightly convex and the glenoid border slopes downward and forward, giving the post-scapular fossa an obscurely triangular shape; it is of nearly the same width as the pre-scapular fossa.   The spine is prominent and is nearly parallel to the glenoid border, ending in a simple, prominent, outwardly projecting acromion, the free border of the spine curving outward to give the acromion prominence.

In the small *H. mustelinus* there is more difference in the scapula than one would expect to find in the same genus and more than appears in any other part of the skeleton. There is a narrower and somewhat more definite coraco-scapular notch and the straight portion of the coracoid border is decidedly longer.   The supra-scapular and glenoid borders are nearly straight, so that, above the neck the blade has an almost rectangular outline and the pre-scapular fossa is somewhat wider than in the larger species.   From the supra-scapular border the spine rises gradually to the highest point and thence remains uniform to the acromion, the level and sloping parts meeting in a more distinct angulation than in the larger species.   The acromion is altogether different from that of *H. cruentus* and consists of a triangular process which is given off from the spine near its distal end and widens to the free border, which is oblique to the course of the spine and that of the shoulder-blade as a whole.   In both larger and smaller species the coracoid is small and incurved.   The scapula of *Canis* is very similar to that of *Hyaenodon*, but the spine pursues a more oblique course, downward and backward across the blade and the acromion differs in shape from that of both the types seen in *Hyaenodon*.

The *pelvis* differs considerably in the larger and smaller species; in the former the ilium is relatively broader and the acetabular border forms a gently concave curvature from the acetabulum to the anterior end; the iliac surface is narrow and bears no distinct pectineal process.   The crest of the ilium is straighter than in *Canis* and forms a sharper angle with the acetabular border.   As in the Fissipedia generally, the gluteal surface is narrow and simply concave, but is relatively narrower in *Hyaenodon* than in *Canis*.   The ischium is short and, in none of the accessible specimens, is there a definite tuberosity; the posterior end is merely roughened.   The pubis is thin and compressed and forms a short symphysis with its fellow; the obturator foramen is relatively large.

In *H. mustelinus* the chief difference from the large species in the pelvis is in the shape of the ilium, which has an angulate acetabular border, the posterior portion forming an obtuse angle with the anterior part; the two ilia also are more strongly everted and the gluteal surface is relatively narrower.   The differences are not at all striking.

LIMBS

As in the Creodonta generally, *Hyaenodon* has a disproportionally large head, which gives a certain grotesque look to the skeleton.   Aside from the head, the limbs are relatively short and the feet weak; except for the muscles of the jaws, the muscular development of these ancient predatory creatures seems to have been less advanced than in most of the existing Fissipedia.

The *humerus* is shaped much as in *Canis*, but somewhat less curved; the head is large and projects well behind the plane of the shaft; the external tuberosity is large and rugose, but low, not rising above the level of the head.   There is no deltoid crest and the line which served for the attachment of the infra-spinatus muscle is faintly or not at all marked.   The shaft is smooth, stout proximally, but tapering rapidly downward; the distal end is broad, but has an inconspicuous supinator ridge.   The internal epicondyle, on the other hand, is very prominent and is perforated by a large foramen; the supra-trochlear fossa in front and the anconeal fossa behind are both deep and are connected by a large opening.   The trochlea is low and has an external convexity and internal concavity for the head of the radius, thus making an interlocking, ginglymoid joint, the radius having lost all power of rotation.

The forearm bones are short and stout.   The *radius* has a transversely extended head, which occupies the whole width of the humeral trochlea, a smooth, curved shaft and a thick distal end.   As compared with the ulna, it is more slender than in *Canis;* the *ulna*, on the other hand, is less reduced.   The olecranon is much longer than in the Recent genus and has a thicker, more rugose end; it does not project backward at all, but is continued proximally in line with the shaft.   The sigmoid notch is deep and the coronoid prominent.   The shaft is heavy and of the ordinary trihedral form, deeply channelled on the external side; the distal end is much thickened.

The fore and hind limbs are, in effect, of nearly equal length, shoulders and rump at approximately the same level, the vertebral centra arch gently upward, but this is compensated for by the neural spines, so that the general appearance of the various species must have been like that of big-headed, short-legged wolves or hyenas, and the small species were more fox-like in proportions.

The *femur* is of nearly the same length as the humerus and is even stouter; the hemispherical head is set upon a long, sharply constricted neck and rises well above the level of the great trochanter.   The latter is low, but broad, massive and rugose, and is so extended antero-posteriorly as to enclose a deep digital fossa, which is bounded below by the rough *linea intertrochanterica;* the second trochanter is more prominent than in *Canis*. In *H. mustelinus* there is a distinct, though very small, third trochanter, of which hardly a vestige remains in the large species and which all Recent Carnivora have lost.   The shaft is stout, cylindrical, nearly straight and very smooth, as it displays none of the *lineae asperae*, which are so conspicuous on the front and hind faces of the femoral shaft in *Canis*.   The rotular trochlea is longer proximo-distally and is more deeply concave than in the Recent genus, but there is no supra-patellar fossa.   The condyles are large and prominent and are separated by a very wide inter-condylar notch.   On the proximal aspect of each condyle is a small articular depression, evidently for the attachment of the *fabellae*, which are not preserved in connection with any of the skeletons.

The *patella* is relatively broader and thicker than it is in *Canis*, but otherwise very much as in that genus, with roughened anterior face, broad proximal and pointed distal end.

The *tibia* is of nearly the same length as the radius; the surface for the external condyle of the femur is larger and reflected farther upon the posterior side than is that for the internal condyle; the spine is very low and inconspicuous.   The external condylar surface projects far out from the shaft and on the distal side of the projection there is an articular surface, which rests upon the head of the fibula.   The cnemial crest is a short rugosity for insertion of the patellar ligament and the extensor muscles of the thigh, but is not continued distally.   The shaft is very straight and of the usual trihedral form, becoming cylindrical distally and broadening at the end; it is stout in the larger species, *H. horridus* and *H. cruentus*, slender and weak in *H. crucians* and *H. mustelinus*.   The distal surface for the astragalus is much more advanced and specialized than in most creodonts; it is divided into inner and outer cavities for the condyles of the astragalus, separated by a low inter-condylar ridge which ends in tongue-like processes on the dorsal and plantar sides, the latter very small.   The internal malleolus is thin transversely, broad antero-posteriorly, long, and bearing on its distal end a facet which fits into a pit on the neck of the astragalus. The whole mechanism of the ankle joint is more fissiped than creodont in character.

The *fibula* is little reduced and is separated from the tibia by a wide inter-osseous space, the two bones touching only at the proximal and distal extremities.   This is in striking contrast to the leg of *Canis*, in which the very slender shaft of the fibula is closely attached to the tibia for more than half its length.   The proximal end of the fibula is heavy and thick, especially in the antero-posterior dimension, and with a deep tendinal sulcus. The shaft is slender, but far less reduced than in *Canis*, remarkably straight and of almost cylindrical shape, a faintly marked crest on the tibial side interrupting the regularity of form.   The distal end is much broadened and thickened to form the heavy external malleolus, which on the outer side has a slightly recurved prominence behind which passed the tendon of the *M. peroneus tertius et brevis;* in *Canis* there is a second prominence and, thus, a groove for the tendon is formed.   On the tibial side of the malleolus is a large plane surface for the side of the astragalar trochlea and on the distal end a facet for the calcaneum. The calcaneo-fibular articulation is almost universal among the Creodonta, while nearly all fissipeds have lost it.

The feet of *Hyaenodon* are relatively weak, with long slender metapodials, and they do not impress one as being particularly well adapted either for swift and sustained running, or for the capture of prey.   They must, however, have been efficient in their own way, for the hyaenodonts successfully bore the brunt of competition for a surprisingly long time in America and in Europe they persisted even later.   Manus and pes are pentadactyl, with unreduced pollex and hallux and divergent metapodials.

MANUS

The *carpus* is very primitive, more insectivore than carnivore in character.   The scaphoid is broad, very short proximo-distally and its articular surface for the radius is continuous with that on the lunar, so closely do the two bones fit together, though there is no sign of coalescence between them.   In both bones the radial surface is carried over so far upon the dorsal side as to suggest a plantigrade gait.   Distally, the scaphoid articulates with the trapezium, trapezoid and central, the latter preventing any contact with the magnum.

The *lunar* is narrower than the scaphoid, but longer proximo-distally and the radial surface is carried over upon the dorsal face even farther than is that on the scaphoid. Distally, the lunar rests upon the central, magnum and unciform.

The *pyramidal* is much the largest bone of the proximal row, especially in transverse width, which is considerably greater on the distal end, where it covers the unciform and projects beyond it as a distinct process.   The proximal articular surface for the ulna is a simple concavity and the entire palmar side of the pyramidal is covered by the facet for the pisiform.

The *pisiform* is a relatively large and massive bone; the proximal end, which articulates with the ulna and the pyramidal, is very broad, contracting abruptly to the free portion, which is narrow, but deep.

In the distal row the largest bone, next after the unciform, is the trapezium, which extends distally beneath the trapezoid and has a large lateral surface for the second metacarpal and a much smaller one for the trapezoid; distally it articulates with mc. I only.

The *trapezoid* is much smaller than the trapezium and has a nearly square dorsal face; distally, it articulates only with mc. II.   The *central* is a small, wedge-like bone, which articulates with four carpals, proximally with the scaphoid and lunar, distally with the trapezoid and magnum.

Notwithstanding its name, the *magnum* is the smallest bone in the carpus, the central only excepted; it also has contact with four carpals, the trapezoid, central, lunar and unciform, fitting into a concavity of the latter; the magnum rests upon the head of mc. III and also has a very limited contact, on the radial side, with mc. II.

The *unciform* is the largest of the carpal elements, especially in transverse width; proximo-distally, the length is greatest on the radial side, diminishing externally; the proximal surface is convex and fits into the distal concavity of the pyramidal.   Distally, it has large surfaces for the heads of mc. IV and V, and a very small, oblique one for mc. III.

The *metacarpus* consists of five fully formed and functional bones, which are relatively short, much more so than the metatarsals.   No two digits are of the same length and in order of length are III, IV, II, V, I.   The *first* (mc. I) is the shortest and most slender of the series; the proximal end is thickened and articulates with the head of mc. I by means of a saddle-shaped surface, which seems to have allowed considerable freedom of motion to the pollex, though probably not to the extent of true opposability; the shaft is slender and the distal trochlea, for articulation with the first phalanx, is hemispherical.   *Metacarpal II* is much longer and stouter than the first; it does not form a symmetrical pair with mc. IV; the actual length of the two bones is not very different, but the head of mc. II extends farther proximally into the carpus than does mc. IV, in consequence of which the *effective* length of the latter is considerably greater.   Mc. II articulates on the radial side with the trapezium, proximally with the trapezoid and, in the ulnar side, it extends slightly over the head of mc. III to a very small contact with the magnum.

Metacarpal III is the longest and, after mc. I, decidedly the most slender of the metacarpals; its proximal end is covered by the magnum and its articulation with the unciform is more lateral than proximal; the shaft is very slender, broadening moderately at the distal end.   *Metacarpal IV* is a little shorter and heavier than mc. III; *mc. V* is much shorter than the three middle ones, mc. II, III and IV, and is but little longer, though decidedly stouter than mc. I.   The metacarpals have a radiating arrangement, as is usual in the five-toed Carnivora.

The *phalanges* have considerable resemblance to those of *Canis*, both in shape and in relative length; those of the proximal row are long and have nearly the same length in the three median digits, II, III and IV, and in digit V, the proximal phalanx is a little shorter and more slender than that of the pollex. The *ungual* phalanges are thick, blunt, slightly decurved and very canine in appearance, but they retain the characteristic feature which is found in almost all creodonts, pinnipeds and insectivores, in being cleft at the tip.

PES

The hind-foot is pentadactyl and, to all appearance, digitigrade. As was seen in connection with the manus, there is some reason to think that the forefoot was plantigrade, but no living carnivore is known in which the manus is plantigrade and the pes digitigrade and no plantigrade pes is known in which the astragalus is so deeply grooved as it is in *Hyaenodon*. The *astragalus*, in this family, has a more deeply grooved trochlea than in any other creodonts except the Mesonychidae of the Eocene; the neck is long and has on its tibial side an articular pit, which receives the facet on the distal end of the internal malleolus. The *calcaneum* is very stout and has on the dorsal side a raised facet for articulation with the fibula, an articulation which almost all creodonts possess, but was early eliminated in the Fissipedia. The distal end of the calcaneum in *Hyaenodon* is broad and somewhat concave and rests upon the slightly convex proximal end of the cuboid. The *tuber calcis* is relatively very heavy, especially in the dorso-plantar diameter, and has a thickened, club-like proximal end, without tendinal sulcus.

The navicular is broad and proximo-distally elongate; its proximal surface is concave and the convex head of the astragalus rests within it, not touching the cuboid. Distally, the navicular has facets for all three cuneiforms. The *ento-cuneiform* is long, narrow and compressed transversely, but thick in the dorso-plantar dimension; it supports the first metatarsal and has a lateral contact with the second. The *meso-cuneiform* is much shorter proximo-distally and has a square, rugose dorsal face; this bone articulates only with the second metatarsal. The *ecto-cuneiform* is much longer proximo-distally and extends well below the level of the meso; it rests upon the head of the third metatarsal and articulates laterally with the second. The *cuboid*, much the largest bone of the distal row of tarsals, articulates proximally only with the calcaneum and has no contact with the astragalus; its principal diameter is the proximo-distal, equalling the combined navicular and ecto-cuneiform.

The five *metatarsals* are, in striking degree, more slender and elongate than the metacarpals; they are arranged with more distinctly mesaxonic symmetry than are the metacarpals, metatarsal III being the longest, and mt. II and IV form a nearly symmetrical pair, but mt. V is much longer than mt. I. The first digit is a well-developed hallux. The tarsal connections of the metatarsals are very simple; mt. I articulates only with the ento-cuneiform; mt. II is wedged in between the ecto- and ento-cuneiforms, with both of which it articulates laterally and with the short meso-cuneiform proximally. Mt. III is in contact only with the ecto-cuneiform, IV and V only with the cuboid. On the external side of the head of mt. V is a prominent knob-like projection, which resembles the same process in *Canis*.

The *phalanges* of the pes are so like those of the manus as to require no particular description.

Species. There are four nominal species of *Hyaenodon*, *H. horridus*, *H. cruentus*, *H. crucians* and *H. mustelinus*, between which the most obvious difference is in size, but there

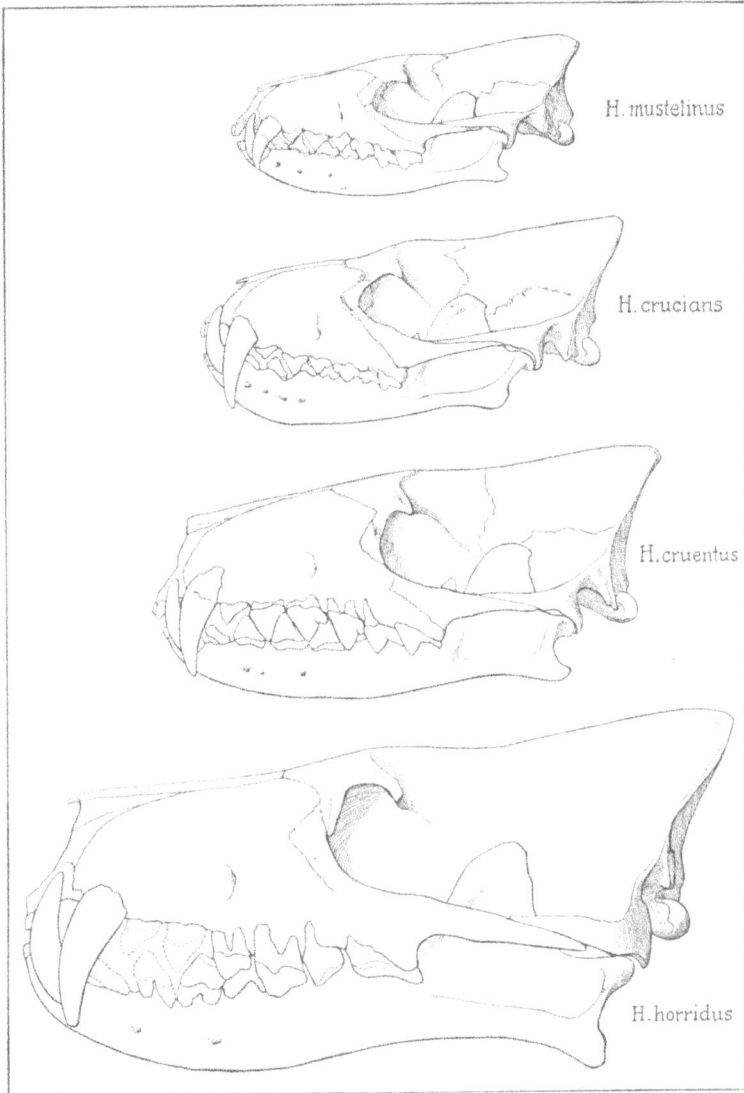

FIG. 4.   Skulls of *Hyaenodon*, drawn to the same scale.   From below upward, *H. horridus*, *H. cruentus*, *H. crucians*, *H. mustelinus*, × 1/5.   Princeton Univ. Mus.

are important structural differences also.   A fifth species, *H. paucidens* Osb. & Wort., is probably also entitled to recognition.   *H. horridus* diverges so strongly from the others that it has been proposed to erect a separate genus, *Neohyaenodon*, for its reception, but, in the opinion of the present writers, this is inadvisable.   Here arises a difficulty so often met with in fossil faunas, the apparent existence at the same time and place of so many species of one and the same genus, an association such as is not to be met with now.   There are several possibilities of removing this difficulty, but, in the lack of evidence, they are merely conjectural and hardly worth discussion.   We may assume, however, that the various species had such different habitats and modes of life that they did not come into direct competition with one another.   It must be remembered that the Bad Land areas were not so much places where the animals lived as gathering grounds for their bones after death and thus creatures from radically different stations, forest and plain, upland and valley, river and swamp, might be assembled in one small area.

Hyaenodon horridus Leidy

(Pl. VII)

Hyaenodon horridus Leidy, Proc. Acad. Nat. Sci. Phila., 1853, p. 292.
Neohyaenodon horridus Thorpe, Amer. Journ. Sci., ser. 5, III, p. 278 (1922).

Leidy gave no formal definition of this species, basing it upon its very large size, "its skull fully equalling that of the largest individuals of the Black Bear, *Ursus americanus*" (Ext. Mamm. Faun. Dak. & Neb., p. 39 (1869)).   He also remarked that in this species the postorbital constriction is very far back.   Osborn and Wortman added, as specific characteristics, the great vertical depth of the side of the face and the presence of an enamel buttress on the outer face of the last lower molar (Bull. Amer. Mus. Nat. Hist., VI, p. 224).   Thorpe, in proposing a new genus for this species, defined it as follows: "Larger than *Hyaenodon*, dolicho-cephalic, glenoids far below basi-cranial plane, basi-cranial region foreshortened, dentition similar to *Hyaenodon*, except for the antero-external buttress on the paraconid of M$\overline{3}$" (*loc. cit.*), but this buttress is not always present.

To these characteristics may be added that the angular process of the mandible is relatively short and that the femur has no third trochanter.   Though the species is thus exceptionally clearly defined, intermediate forms, transitional to *H. cruentus*, are not lacking.   These smaller individuals have been interpreted, perhaps correctly, as females of *H. horridus*, though secondary sex differences are seldom conspicuous in the predaceous mammals.   All the species of this genus and, indeed, almost all creodonts are remarkable for their disproportionately large heads, but *H. horridus*, in which the head nearly equalled that of a Grizzly Bear, had a body and limbs not greatly exceeding those of a Grey Wolf and must have had a grotesque appearance.

In the following table, the dimensions of the skull and upper teeth are taken from a very large animal (Princeton Mus. No. 12,656), but, as the jaws are closed and the lower teeth concealed by the overlapping superior series, the measurements of the lower teeth are taken from an isolated mandible (Princeton Mus. No. 12,649), which is of almost exactly the same size as the lower jaw attached to the skull No. 12,656.

MEASUREMENTS

Upper incisor series, width	32.0 mm.	Lower p4, ant.-post. diam	21.0 mm.
Upper canine, ant.-post. diam	18.0	Lower m1, ant.-post. diam	13.0
Upper p1, ant.-post. diam	21.0	Lower m2, ant.-post. diam	19.0
Upper p2, ant.-post. diam	22.0	Lower m3, ant.-post. diam	30.0
Upper p3, ant.-post. diam	20.0	Skull, median basal length	315.0
Upper p4, ant.-post. diam	26.0	Skull, length i1 to occ. cond. incl	331.0
Upper m1, ant.-post. diam	19.0	Skull, extreme length	337.0
Upper m2, ant.-post. diam	33.0	Occiput, width at base	108.0
Upper cheek-teeth series, length	138.0	Occiput, height (est.)	107.0
Upper premolar series, length	92.0	Cranium, width at post-orb. constr.	45.0
Upper molar series, length	51.0	Face, width over post-orb. proc	118.0
Lower canine, ant.-post. diam	19.5	Mandible, length c̄ to cond. incl	276.0
Lower canine, transverse diam	17.0	Mandible, length c̄ to angle incl	272.0
Lower cheek-teeth series, length	143.0	Mandible, height of condyle	48.0
Lower premolar series, length	77.0	Mandible, height of coronoid	94.0
Lower molar series, length	68.0	Mandible, depth at m3̄	53.0
Lower p2, ant.-post. diam	15.0	Mandible, length of symphysis	109.0
Lower p3, ant.-post. diam	19.0		

Hyaenodon cruentus Leidy

(Pl. IX, Figs. 1–5; Pl. XXII, Figs. 5, 6)

Hyaenodon cruentus Leidy, Proc. Acad. Nat. Sci. Phila., 1853, p. 393.

Leidy's original definition gave only comparative dimensions; the species "was between a fourth and a third less than *H. horridus* and was rather larger than the *H. brachyrhynchus* of France"; in addition, his figure showed that the buttress seen on the outer side of the anterior cusp of m3̄ in *H. horridus* was not present. The difference of size, "between a third and a fourth," is within the limits of individual variation and the buttress is not always present even in the largest individuals. *H. cruentus* agrees with *H. horridus* and differs from the smaller species (1) in the position of the postorbital constriction, (2) in the remarkable preorbital height of the face, (3) in large size. There are, however, certain structural differences between the two larger species, if they are distinct species; (1) the postorbital constriction is farther forward in some individuals, and (2) the angular process of the mandible is much better developed in the typical examples of *H. cruentus* than in those of *H. horridus*, but the constancy of this difference remains to be proved.

We are inclined to doubt that the two alleged species are anything more than larger and smaller races of one species. There is much variation in size of these larger hyaenodonts and it is probable that a sufficiently numerous suite of skulls would give all the transitions in size between one supposed species and the other. Possible sex differences in size must also be borne in mind; the fine skeleton in the American Museum (No. 1375) labelled "*Hyaenodon horridus*, female," is a little larger than a skeleton in the Princeton Museum (No. 10,995) which is referred to *H. cruentus;* but, otherwise, they are very similar.

On the other hand, *H. cruentus* has been recognized as distinct for more than eighty years and it seems best to retain it until its identity with *H. horridus* can be proved.

In the subjoined tables A.M.N.H. No. 1375 is the skeleton labelled "*H. horridus*, female." Princeton No. 10,995 is the skeleton in the Princeton museum referred to *H. cruentus*. Princeton No. 10,010 is an isolated skull in the same collection and is a typical example of *H. cruentus*. For purposes of comparison the dimensions of the Princeton skeleton of *H. mustelinus* (No. 13,583) are added.

MEASUREMENTS

	Hyaenodon mustelinus	Princeton No. 10,010	Princeton No. 10,995	A.M.N.H. No. 1375
	mm.	mm.	mm.	mm.
Upper incisor series, width.................	17.0	30.0	29.0	38.0
Upper canine, ant.-post. diam..............	9.0	12.0	16.0	18.0
Upper canine, transverse diam..............	5.0	9.0	12.0	14.0
Upper p1, ant.-post. diam..................	8.0		12.0	12.0
Upper p2, ant.-post. diam..................	10.0	14.5	16.0	19.5
Upper p3, ant.-post. diam..................	10.0	18.0	17.5	18.0
Upper p4, ant.-post. diam..................	10.0	18.0	19.0	18.0
Upper m1, ant.-post. diam.	10.0	16.0	15.0	16.0
Upper m2, ant.-post. diam.................	13.0	23.0	27.0	28.0
Upper cheek-teeth series, length............	67.5	103.0	107.0	114.0
Upper premolar series, length..............	49.0	63.0	68.0	71.5
Upper molar series, length.................	23.0	40.0	38.0	42.0
Lower canine, ant.-post. diam..............	9.0	15.0		
Lower canine, transverse diam..............	6.0	11.0		
Lower cheek-teeth series, length............	71.0	112.0		129.0
Lower premolar series, length..............		58.0		75.0
Lower molar series, length.................		54.0		52.0
Lower p1, ant.-post. diam..................				12.0
Lower p2, ant.-post. diam.................		13.0		18.0
Lower p3, ant.-post. diam..................				19.0
Lower p4, ant.-post. diam..................		17.0		20.0
Lower m1, ant.-post. diam.................		14.0		14.0
Lower m2, ant.-post. diam.................		17.0		18.0
Lower m3, ant.-post. diam.................		24.0		25.0
Skull, median basal length.................	152.0	213.0	207.0	254.0
Skull, length i1 to occ. cond. incl............	158.0	226.0	230.0	259.0
Skull, extreme length.....................	164.0			
Occiput, width at base....................	45.0	78.0	65.0	87.0
Occiput, height fr. basi-occ................	46.0		70.0	86.0
Cranium, width at post-orb. constr..........	20.0	38.0	35.0	38.0
Cranium, width over post-orb. proc..........	56.0	77.0	82.0	95.0
Palate, width at m2......................	40.0	70.0		68.0
Palate, width at p1	15.0	23.0		24.0
Mandible, length c̄ to cond. incl.............	115.0	189.0	199.0	233.0
Mandible, length c̄ to angle incl.............	121.0		202.0	226.0
Mandible, height of cond. fr. vent. bord.......	14.0	32.0	34.0	34.0
Mandible, height of coronoid...............	39.5		60.0	72.0
Mandible, depth at m3̄....................	12.0	33.0		40.0
Mandible, length of symphysis.............	38.0	69.0	71.0	75.0
Neck, length............................	97.0		150.0	182.0
Atlas, ant.-post. width of neur. arch........	9.5		13.0	11.0
Atlas, width over post. cotyles.............			42.0	44.0
Axis, length of centrum...................	17.0		24.0	38.0
Axis, ant.-post. width of neur. spine........	34.0		53.0	64.0
First dorsal, height of spine...............	32.0			23.0
First lumbar, length of centrum............	18.0			30.0

MEASUREMENTS (*Continued*)

	Hyaenodon mustelinus	Princeton *No. 10,010*	Princeton *No. 10,995*	A.M.N.H. *No. 1375*
	mm.	mm.	mm.	mm.
Second lumbar, length of centrum...........	18.0			32.0
Third lumbar, length of centrum...........	18.0			33.0
Sacrum, length...........................	31.0			67.0
First caudal, length......................	11.0			17.0
First caudal, width over trans. proc.........	30.0			42.0
Fifth caudal, length......................	10.0			16.0
Fifth caudal, width over trans. proc.........	10.0			28.0
Tenth caudal, length.....................	11.5			21.0
Tenth caudal, width over trans. proc........				9.0
Fifteenth caudal, length..................	19.0			22.0
Fifteenth caudal, width over trans. proc......				8.0
Twentieth caudal, length.................	16.0			10.0
Twentieth caudal, width over trans. proc.....				5.0
Twenty-fifth caudal, length...............	16.0			10.0
Twenty-fifth caudal, width................				2.0
Scapula, prox.-dist. length...............	78.0		131.0	130.0
Scapula, greatest width...................	41.0		69.0	87.0
Humerus, length from head................	96.0		138.0	172.0
Humerus, width of dist. end...............	28.0		38.0	43.0
Ulna, length.............................				?191.0
Ulna, length fr. coron. proc................				?170.0
Ulna, width across hum. surf..............	14.0			23.0
Radius, length on ulnar side..............				151.0
Radius, width of prox. end.................	14.0		20.0	23.0
Radius, width of dist. end.................			18.0	27.0
Metacarpal I, length.....................				35.0
Metacarpal I, width of prox. end...........				12.0
Metacarpal II, length....................				57.0
Metacarpal II, width of prox. end..........				13.0
Metacarpal II, width of dist. end...........				10.0
Metacarpal III, length....................				68.0
Metacarpal III, width of prox. end.........				11.0
Metacarpal III, width of dist. end..........				11.0
Metacarpal IV, length....................				60.0
Metacarpal IV, width of prox. end..........				12.0
Metacarpal IV, width of dist. end..........				11.0
Metacarpal V, length				39.0
Metacarpal V, width prox. end				12.0
Metacarpal V, width dist. end				11.0
Pelvis, length...........................	114.0			192.0
Pelvis, width over ilia....................	69.0			98.0
Pelvis, width over ischia..................	50.0			95.0
Pelvis, length of symphysis...............				45.0
Ilium, length............................	64.0			120.0
Ilium, greatest width.....................	19.0			25.0
Ischium, length..........................	43.0		65.0	68.0

MEASUREMENTS (*Continued*)

	Hyaenodon mustelinus	Princeton No. 10,010	Princeton No. 10,995	A.M.N.H. No. 1375
	mm.	mm.	mm.	mm.
Femur, length, head to int. cond............	108.0		168.0	195.0
Femur, gr't troch. to ext. cond..............	104.0		161.0	196.0
Femur, width of prox. end..................	25.0		38.0	46.0
Femur, width of dist. end..................	24.0		33.0	43.0
Tibia, length on int. side..................	109.0		146.0	179.0
Tibia, width of prox. end..................	25.0		40.0	41.0
Tibia, width of dist. end..................	12.0		26.0	20.0
Fibula, length............................				167.0
Fibula, width of prox. end.................				15.0
Fibula, width of dist. end.................				12.0
Calcaneum, length........................	? 31.0		50.0	57.0
Astragalus, width over trochlea.............	10.0		15.0	15.0
Metatarsal I, length......................			41.0	39.0
Metatarsal I, width of prox. end............			6.0	9.5
Metatarsal I, width of dist. end............			11.0	8.0
Metatarsal II, length.....................			59.0	60.0
Metatarsal II, width of prox. end...........			8.0	11.0
Metatarsal II, width of dist. end...........			8.0	10.0
Metatarsal III, length....................			65.0	68.0
Metatarsal III, width of prox. end..........			8.0	10.0
Metatarsal III, width of dist. end..........			9.0	10.0
Metatarsal IV, length.....................			64.0	68.0
Metatarsal IV, width of prox. end..........			8.0	9.5
Metatarsal IV, width of dist. end..........			9.0	9.0
Metatarsal V, length......................			48.0	50.0
Metatarsal V, width of prox. end...........				12.0
Metatarsal V, width of dist. end...........			7.0	8.0
Manus, dig. III, 1st phalanx, length.......				26.0
Manus, dig. III, 2nd phalanx, length........				14.0
Manus, dig. III, ungual, length.............				18.0
Pes, dig. III, 1st phalanx, length...........			25.0	25.0
Pes, dig. III, 2nd phalanx, length..........			11.0	12.0
Pes, dig. III, ungual, length...............			13.0	19.0

Horizon: Chadron & Brulé.

Hyaenodon crucians Leidy

Hyaenodon crucians Leidy, Proc. Acad. Nat. Sci. Phila., 1853, p. 393.
Hyaenodon leptocephalus Scott, Journ. Acad. Nat. Sci. Phila., IX, p. 175 (1888).

There can be no doubt as to the distinctness of this species, which is not only constantly much smaller than the two preceding ones, but also has a strikingly different looking skull. In part, this difference is in the much shorter distance between the eye-sockets and the postorbital constriction and partly in the shape of the occiput, the dorsal portion of which is much broader, not lanceolate.

H. leptocephalus was proposed because of the union in the median ventral line of the

˙pterygoid plates of the alisphenoids in the type-skull, but this characteristic seems to be individual rather than specific.

In length of skull, *H. crucians* slightly exceeds that of the Coyote.

In the Princeton collection there is a very perfect skull with mandible in place (No. 12,954) and the skull and part of a skeleton of a young animal with milk-teeth (No. 10,916).

MEASUREMENTS

	No. 12,954	No. 10,916
Upper incisor series, width	20.0 mm.	mm.
Upper canine, ant.-post. diam	11.0	
Upper canine, transv. diam	7.0	
Upper cheek-teeth series, length	76.0	
Upper premolar series, length	50.0	
Upper molar series, length	25.0	
Upper p1, ant.-post. diam	10.0	
Upper p2, ant.-post. diam	13.0	
Upper p3, ant.-post. diam	13.0	
Upper p4, ant.-post. diam	12.0	
Upper m1, ant.-post. diam	10.0	18.0
Upper m2, ant.-post. diam	16.0	
Upper milk-canine, ant.-post. diam		11.0
Upper milk-canine, transverse diam		9.5
Upper dp2, length		13.0
Upper dp3, length		16.0
Upper dp4, length		18.0
Lower canine, ant.-post. diam	9.0	
Lower canine, transverse diam	8.0	
Lower milk c. ant.-post. diam		11.0
Lower milk c. transverse diam		7.0
Lower p1, length	9.0	14.0
Lower p2, length	10.5	12.0
Lower dp 3, length		14.0
Lower dp4, length		14.0
Lower m1, length		13.0
Lower m 2, length		17.0
Skull, median basal length	188.0	
Skull, length i1 to occ. cond. incl	191.0	187.0
Skull, extreme length	206.0	
Skull, occiput, width at base	58.0	55.0
Skull, occiput, height above basi-occ	68.0	
Cranium, width at postorb. constr	28.0	32.0
Cranium, width over postorb. proc	70.0	70.0
Mandible, length c̄ to cond. incl	161.0	162.0
Mandible, height of cond. fr. vent. border	33.0	
Mandible, height of coron. fr. vent. border	55.0	
Mandible, depth at m2̄	27.0	29.0
Mandible, length of symphysis	56.0	61.0

An interesting fact is brought out by the young animal with milk-teeth (No. 10,916), of which the measurements are given in the preceding table and that is the very early eruption of the first upper and first and second lower molars. As has long been noted, the first molars, especially in the lower jaw, of a full-grown *Hyaenodon* are remarkably small and almost seem to be in a state of atrophy. As Wortman was the first to point out, this appearance is due to the long time of wear that these teeth have undergone, longer than for any other teeth. When freshly erupted, the teeth in question do not look unduly small and are not in any way degenerate. They were, however, in full use before any other permanent teeth were in place and, from the beginning, they were subjected to more intensive wear.

Among the skeletal material obtained in connection with the young skull is a well-preserved tibia and the most perfect fibula yet found. These were described in the account of the genus and it remains only to give their dimensions.

MEASUREMENTS

Tibia, length, inner side..............	132.0 mm.	Fibula, length.....................	123.0 mm.
Tibia, width prox. end..............	33.0	Fibula, width prox. end..............	12.0
Tibia, thickness prox. end...........	21.0	Fibula, thickness prox. end..........	14.0
Tibia, width dist. end...............	18.0	Fibula, width dist. end..............	8.0
Tibia, thickness dist. end...........	19.0	Fibula, thickness dist. end..........	13.0

Horizon: the species occurs in all parts of the White River series, except in the *Leptauchenia-Protoceras* Beds.

Hyaenodon paucidens Osb. & Wort.

Hyaenodon paucidens Osborn & Wortman, Bull. Amer. Mus. Nat. Hist., VI, p. 223 (1894).

Characterized by the absence of the first upper premolar, giving the formula: $p\frac{3}{4}$, $m\frac{2}{3}$.

The authors say: "The absence of this tooth is not an accidental variation. In the first place, the space which the first premolar should occupy is relatively shorter than in the nearest ally, *H. crucians*, being only 10 mm., whereas in *H. crucians* it is 15 mm. The entire length of the tooth line measuring from the posterior border of the upper canine to the posterior border of the last molar is 80 mm. in *H. crucians* and 70 mm. in *H. paucidens*. The third premolars in both upper and lower jaws have a more oblique position and the teeth are more crowded than in *H. crucians*." *H. paucidens* agrees with the larger species of the genus in having the postorbital constriction at the fronto-parietal suture. This species is the only known American representative of the short-faced hyaenodonts found in Europe.

Horizon: not stated.

Hyaenodon montanus Douglass

Hyaenodon montanus Douglass, Trans. Amer. Phil. Soc. (2), XX, p. 243.

This species is not improbably distinct, but it is still so incompletely known, that its status is uncertain.

In the Princeton collection there is a *Hyaenodon* skull from the upper part of the Chadron beds on Indian Creek, South Dakota (No. 12,790), which seems referable to this species. In size, this skull is intermediate between *H. crucians* and *H. cruentus*, but differs from both of those species in the greater relative breadth of the face and palate, which gives an unusual expression of sturdiness to the head. The pterygoids and posterior part of the palatines are separated more widely and the slit between them is carried farther forward than in any of the other species and to an extent that does not seem due to individual variability.   As the premaxillaries have been lost, the full length of the skull cannot be ascertained.

MEASUREMENTS

Skull, length c to occ. cond..........	190.0 mm.	Upper canine, ant.-post. diam.........	13.0 mm.
Skull, width at post-orb. constr.......	30.0	P4, ant.-post. length.................	15.0
Skull, width over post-orb. proc......	76.0	M1, ant.-post. length................	13.0
Occiput, width at base..............	58.0	M2, ant.-post. length...............	18.0

Horizon: Chadron.

Hyaenodon mustelinus Scott

(Pl. VIII, Pl. IX, Figs. 6–11, Pl. XVI)

Hyaenodon mustelinus Scott, Journ. Acad. Nat. Sci. Phila., IX, p. 499 (1894).

This is not only the smallest of the American species of the genus, but is further distinguished by a number of structural characters.   It has a much longer tail proportionately than the larger species; the scapula has a remarkable T-shaped acromion and the femur has a distinct third trochanter.   The postorbital constriction is, like that of *H. crucians*, placed close behind the orbits.   The anterior ends of the nasals have simple median points and are not notched.

The measurements are given in the table with those of *H. cruentus*.

Horizon: this rare species has been found only in the lower Brulé.

Hyaenodont Incertae Sedis

Under the name of *Hyaenodon minutus* Douglass described a small creodont from the Chadron of Montana, and Matthew (Bull. Amer. Mus. Nat. Hist., XIX, p. 208) referred to the same species an upper jaw-fragment from the same horizon and locality and tentatively assigned it to Schlosser's genus *?Pseudopterodon*, which was established on milk-teeth of *Hyaenodon*.   The fragment is of interest as probably representing the survival of some Eocene member of the family, but is too incomplete for the founding of a new genus on it.

Suborder **FISSIPEDIA** Fischer de Waldheim

The classification and subdivision of the non-marine Carnivora have long been subjects of discussion and the final word has yet to be spoken.   To English-speaking naturalists Flower's classical paper (Proc. Zool. Soc. Lond., 1869, p. 4) long remained a satisfactory solution of the problem, though it never was entirely acceptable to palaeontologists, especially in Germany.   Flower divided the Fissipedia into three "primary sections" (superfamilies, they would be called now), which he named Arctoidea, Cynoidea and Aeluroidea.   Winge (1895) united the Cynoidea and Arctoidea into one section, to which he gave the latter name.   This manifest improvement was adopted by M. Weber in his monumental *Säugethiere*.   Winge (1895) retained the group Aeluroidea, changing the name to Herpestoidea.   Long ago Schlosser emphasized the isolation of the cats, which he separated widely from all other Carnivora, though without making a formal arrangement of superfamilies.   As a result of our studies of North American Tertiary cats, we are inclined to follow Schlosser's lead (Biolog. Centralblatt, Bd. VIII, pp. 587–600, 609–31) and further, to do what he refrained from doing and tentatively erect a superfamily for the cats alone.

The arrangement, modified from Flower and Winge is in three superfamilies, for two of which Flower's terms are employed: I Cynoidea, including the Arctoidea and Cynoidea of Flower; II Herpestoidea, including the viverrines and hyenas, and III *Aeluroidea*, containing the cats only.   It is a question whether it is more objectionable to make use of Flower's and Winge's terms in senses different from those for which they were originally proposed, or to coin new names.   We prefer to keep the familiar terms as the less of two evils.

While we cannot but agree with Winge in merging two of Flower's "primary sections," we think it better to use the term Cynoidea, rather than Arctoidea, for the dogs represent the main line of carnivorous descent, while the bears are a specialized side-branch that arose late in the history of the suborder.

FISSIPEDIA

Osborn 1921	Hay 1930	Simpson 1931	Scott & Jepsen 1936
Arctoidea	Ursoidae	Canoidea	Cynoidea
Canidae	Canidae	Canidae	Canidae
Procyonidae	Procyonidae	Procyonidae	Procyonidae
Ursidae	Bassariscidae	Ursidae	Ursidae
Mustelidae	Ursidae	Mustelidae	Mustelidae
	Mustelidae		
Aeluroidea	Feloidae	Feloidea	Herpestoidea
Viverridae		Viverridae	Viverridae
Protelidae		Hyaenidae	Hyaenidae
Hyaenidae	Hyaenidae		Aeluroidea
Felidae	Felidae	Felidae	Felidae

Flower 1869	Winge 1895	S & J 1936
Arctoidea Cynoidea }=	Arctoidea =	Cynoidea

$$\text{Aeluroidea} = \text{Herpestoidea} = \begin{cases} \text{Herpestoidea} \\ \text{Aeluroidea} \end{cases}$$

In comparison with the Oligocene of Europe and Asia, the White River beds have yielded but a scanty list of fissipeds and much the greater part of these belong to the Canidae and Felidae.   The Mustelidae are very rare and show little diversification; the civets (Viverridae) and hyenas, so far as is known, never invaded the Western Hemisphere at all, while the bears (Ursidae) and raccoons (Procyonidae) would seem not to have been in separate existence so early as the Oligocene; at least, nothing has been found, either in the New World, or the Old, before the Miocene, which could be referred to these groups, though supposed ancestors are known.

Of the dog family (Canidae) there are in the White River two well-distinguished series or tribes, one of larger animals, typified by *Daphoenus*, the other consisting of very much smaller forms, of which the commonest is *Pseudocynodictis*.   Not only is the latter distinguished by small size, but also by the auditory bullae, which are large and firmly attached to the skull, while in *Daphoenus* the bullae are small and so loosely connected, that they are very rarely found.

Superfamily I. **CYNOIDEA**

Family 1. CANIDAE

The two phylogenetic series, or tribes, of this family, mentioned above, are traceable back to the upper Eocene, when they were already separate and distinct, and they would seem to have arisen from a common ancestry in the middle Eocene, but the known Bridger fossils which are supposed to be referable to this family are too fragmentary to justify positive statements. On the other hand, the upper Eocene dogs of North America may have been immigrants from Asia.

Daphoenus Leidy

(Pls. II, X, XI, XII)

Daphoenus Leidy, Proc. Acad. Nat. Sci. Phila., 1853, p. 393.
Amphicyon Leidy (*nec* Lartet), *ibid.*, 1854, p. 57.
Canis Cope (*nec*. Linn.), Ann. Rep. U. S. Geolog. Surv. Terrs., 1873, p. 505.
Proamphicyon Hatcher, Mem. Carn. Mus. Pittsburgh, I, p. 95.

This genus is very well known, thanks to Hatcher's discovery of the almost complete skeleton now in the Carnegie Museum at Pittsburgh and the fine series of skulls in the Princeton museum, collected by the late Professor W. J. Sinclair, with additional material in the American and Field Museums.

DENTITION

The dental formula is unreduced and reads: $i\frac{3}{3}$, $c\frac{1}{1}$, $p\frac{4}{4}$, $m\frac{3}{3}$; All the teeth are characteristically "microdont" canine in form, a form from which the teeth of all the other fissiped families might be derived and, perhaps, were actually so derived. Not that American genera were ancestral to all the families of the Fissipedia; they certainly were not, but *Daphoenus* retained the *kind* of teeth which might very well have been ancestral to the other groups and which still persists, with surprisingly little change, in most of the wolves, jackals and foxes of the present time.

Upper Teeth. The *incisors* are small and are arranged in a straighter transverse line than are those of *Canis;* as in the latter i3 is the largest of the series, but has a long, simple crown and is relatively smaller than in the modern genus. The *canines* are stout fangs and vary considerably in degree of lateral compression, but are not in any way unusual.

The *premolars* are relatively small, much more so than in typical species of *Canis*, and are well spaced apart. The reduction of the premolars is most conspicuous in *D. nebrascensis*, which Hatcher made the type of his *Proamphicyon;* in shape they resemble those of *Canis*, but are somewhat simpler, usually having no basal cusps. P1 and 2 are implanted by two roots, p3 sometimes has three, as we have observed in one specimen of *D. vetus* and one of *D. hartshornianus*, but this feature is not constant. The sectorial (p4) is decidedly microdont, in Huxley's sense of the term; the posterior cusp of the shearing blade (tritocone), being small, but the antero-internal cusp (deuterocone) is much more distinct than in *Canis* and has rather the appearance found in such existing South American genera as *Cerdocyon* and *Chrysocyon*, the dentition of which has certain striking resemblances to that of the North American *Urocyon*.

The *molars* diminish rapidly in size from m$\underline{1}$ to m$\underline{3}$; the first (m$\underline{1}$) is triangular, with two external conical cusps and a single, internal, crescentic one, on the horns of which near the outer side, are minute conules, the anterior one being hardly visible.   The cingulum is prominent, especially on the inner side, where it forms a broad shelf.   M$\underline{2}$ is similar in form to m$\underline{1}$, but much smaller, especially in *D. hartshornianus;* it differs from m$\underline{1}$ in the reduction of the postero-external cusp, which gives the crown an asymmetrical shape.   The third molar is usually lost from the skull, though the open alveoli show that the loss did not occur long before death.   When present, m$\underline{3}$ has a very small, transversely oval crown, which is usually implanted by two roots, but sometimes by a single one.   The two external cones and the inner crescent and cingulum are faintly indicated and are soon obliterated by wear.   The tooth is decidedly more reduced in *D. hartshornianus* than in *D. vetus*, but is always visible externally and is not overlapped and concealed by m$\underline{2}$.

Lower Teeth.   The *incisors* are very small and are usually lost from the jaw; none of those at hand are so well preserved as to show the bifid crown, if such were originally present.

The *canines* are of the usual, dog-like character.   The *premolars* are very much like those of the modern members of the family, among which there are but slight differences of shape; in some genera these teeth are higher, thinner, more compressed and more sharply pointed than in others, and there is also a limited degree of variation in the development of the cingulum and accessory basal cusps on p$\overline{3}$ and $\overline{4}$.   There are also small differences in these respects between the larger and smaller species of *Daphoenus*, as will be shown in the descriptions of species.

The first premolar (p$\overline{1}$) is very small, though less reduced than in some Recent genera; P$\overline{2}$, $\overline{3}$ and $\overline{4}$ increase regularly in size posteriorly; they are all thin, laterally compressed and sharp-pointed cones, inserted by two roots.   The cingulum is indistinct, except on the hinder side, especially of p$\overline{4}$.   A very minute accessory cusp appears on the hinder margin of p$\overline{3}$ and a very distinct one on p$\overline{4}$, which, however, is relatively smaller than in *Canis*, but more prominent than in most other Recent genera, such as *Urocyon, Cerdocyon, Icticyon* (in which it is lacking); in *Otocyon* all the premolars are extremely simple and reduced in size. In the more ancient and primitive *Daphoenus dodgei*, from the Chadron substage, the inferior premolars are lower, thicker and less acutely pointed and the posterior basal cusps are relatively larger.

The *molars* are very much like those of the modern members of the family; the lower sectorial (m$\overline{1}$) has a high anterior, shearing triangle, or "trigonid," the cusps of which are reminiscent of those seen in the primitive Eocene family of the Uintacyonidae, the inner cusp being larger and more distinct than in most existing dogs, but not more than in some Recent genera, such as *Urocyon;* the heel is basin-like, enclosed by internal and external ridges, which, in entirely unworn teeth are finely tuberculate; posteriorly, the two ridges do not quite meet, but leave a narrow groove between them.   In the modern dogs there is great variety in the minutiae of the heel; in most genera it is basin-shaped, but in *Cyon* and *Icticyon* it is a single trenchant cusp.

The second molar (m$\overline{2}$) is made up very much as is m$\overline{1}$, with anterior triangle and posterior basin, but the triangle is very low and not shearing and the tooth is effectively tubercular.   The cusps are very much as in *Canis* and, therefore, not nearly so high, or sharp-pointed as in *Urocyon*, or *Cerdocyon*.   The third molar (m$\overline{3}$) is relatively larger and less reduced than in the modern genera, except *Otocyon*.   Though very small, it retains vestiges of the anterior triangle and the heel, but these are visible only in entirely unworn teeth.

In general appearance the skull is unmistakably canine, but it retains many primitive features from its Eocene ancestry. As compared with the skull of *Canis*, the cranium is long and the face short and the short muzzle is much more abruptly narrowed in front of the eyes. The brain-case is decidedly narrower and of smaller capacity, in consequence of which the sagittal and occipital crests are very much more prominent than in any existing member of the family. The postorbital constriction, which indicates the anterior boundary of the cerebral fossa, is much farther behind the eyes than in *Canis* or *Vulpes* and the brain-case tapers anteriorly much more. The cerebellar fossa, on the other hand, is relatively much longer than in the Recent genera and the glenoid cavities have a correspondingly more anterior position. On account of the very long cranial region, the zygomatic arches are more elongate and somewhat heavier than in modern dogs. The upper contour of the skull is nearly straight, the forward descent of the forehead being slight and gradual.

The *occiput* varies in shape in different individuals, being considerably wider in some than in others, and the differences in shape are sometimes so marked, as to suggest a specific distinction. In the largest skulls, the occiput is wider and the crest describes a more open curve than in those of somewhat smaller size, but in all of them the occiput is narrower at the base, wider at the top and less like a Gothic arch than in the modern Canidae. The paroccipital processes are short and blunt and have a backward inclination, so that they are widely separated from the small auditory bullae. In *Canis*, on the contrary, there is a narrow, sutural contact between the bullae and the paroccipitals.

While the whole aspect of the skull of *Daphoenus*, despite its creodont-like proportions of cranial and facial regions, is unmistakably canine, there are many differences from the skulls of Recent dogs. Many points of detail in the cranium are, however, indeterminable, notwithstanding the fine series of skulls at our disposal, because of the obliteration of the sutures, even in fairly young animals with unworn teeth. Many skulls, otherwise in beautiful preservation, and entirely free from crushing, are yet so extensively cracked, that the sutures are no longer recognizable. Nevertheless, many structural details may be made out which are of evolutionary significance.

The *basioccipital* is much broader proportionately than in the modern genera, because of the very small size of the auditory bullae; its ventral surface is nearly flat and, in most skulls, there is no indication of a keel. The whole *basis cranii*, in correspondence with the elongate cerebellum and medulla oblongata, is relatively much more prolonged behind the glenoid cavities, than it is in *Canis*. The tympanic bullae were evidently very loosely attached to the skull, as they are almost always missing. Among the many individuals which we have examined, there is only one in which these bones remain in their original positions. The tympanics are small and occupy less than half of the fossae in which the petrosal is embedded. Small as it is, the bone is yet inflated and hollow and the auditory meatus has no tubular prolongation, not even such a short one as occurs in *Canis*. Very probably, the remainder of the periotic fossa was occupied by a cartilaginous entotympanic, which did not ossify. This is a much more primitive condition of this region than is to be found in any existing carnivore. The mastoid portion of the periotic is less exposed on the surface than it is in *Canis* and is limited to the lower part of the space between the squamosal and the ex-occipital; the mastoid process is a small rugosity, much as it is in *Canis*.

The *parietals* are very large and form a greater proportion of the cranial roof than they do in the modern genera of the family. The much smaller relative size of the cerebral hemispheres is shown in the shape of the brain-case, which has a large and deep concavity behind the vaulted cranial roof; in the modern dogs there is hardly a remnant of this concavity, the convex vault reaching almost to the occipital crest. The squamosal differs from that of *Canis* only in minor details, one of which is a broad, shelf-like projection over the auditory meatus, which is much narrower in *Canis*, no doubt, because of the more capacious brain-case in the latter. The zygomatic process and the jugal require no particular description except to note that the postorbital angulation of the latter is even less prominent than in Recent species. The *lachrymal* is small and has, when intact, a prominent spine, which has been broken away in most of the fossils; the lachrymal foramen opens within the orbit.

The *frontals* are proportionally smaller than in the modern genera of the family and enclose but little of the cerebral fossa. Hardly any of the sagittal crest is on the frontals and the temporal ridges, which diverge anteriorly from the crest, differ somewhat in the various skulls. The material is not yet sufficient to show whether these differences are specific, or individual, more probably the latter. In the larger skulls the ridges diverge more abruptly and in the smaller ones more gradually. The postorbital processes of the frontals, like those of the jugals, are smaller than in *Canis*. None of the available skulls is so broken as to expose the sinuses, but the convexities of the forehead seem to indicate their presence, though those convexities are not a sure proof of the existence of frontal sinuses, as *Urocyon* demonstrates.

The *nasals* differ little from those of *Canis*, except in being relatively shorter. The *premaxillaries*, on the other hand, differ decidedly from those of the modern genus in being much smaller and especially narrower; the alveolar portion is constricted in front of the canine, forming a groove into which the lower canine is received, much as in the Red Fox (*Vulpes*). The ascending rami are very slender, but long, almost reaching the nasal processes of the frontals.

The *maxillaries*, especially the preorbital portion, are much shorter than in *Canis*, in consequence of which the infraorbital foramen is brought nearer to the orbit and the preorbital constriction of the rostrum is much more abrupt. The palate is like that of *Canis*, though somewhat shorter and wider. The *palatines* are relatively longer than in the modern genera and the posterior nares open farther back. In none of the skulls are the limits of the pterygoids clearly displayed. The cranial foramina are like those of *Canis*, including the presence of an alisphenoid canal.

The *mandible* is very wolf-like, but is, in some respects, more primitive than in existing genera. The horizontal ramus is stouter and the angular hook is less raised above the ventral border; there is no sub-angular lobe such as occurs in *Urocyon* and other Recent genera. The ascending ramus has greater antero-posterior breadth, the coronoid is more inclined backward than in *Canis* and the masseteric fossa is deeply impressed. The condyle is relatively longer than in modern dogs and farther removed from the third molar. The great breadth of the ascending ramus is a primitive creodont-like feature and is to be correlated with the elongation of the cranium and zygomatic arches.

From the height of the occipital and sagittal crests, the size and depth of the masseteric fossa and the stoutness of the zygomatic arch, it may be inferred that the animal had an unusual development of the jaw-muscles, the temporal muscles, in particular, must have been far larger and more powerful than in existing wolves and the whole apparatus of the jaws must have been most formidable, though, as will be shown in what follows, the limbs and feet do not indicate that these early dogs could have had such powers of sustained speed as do their modern successors.

The *Brain* characters are incompletely displayed in a natural cast, which permits comparison with that of *Cynodesmus,* of the lower Miocene. The cerebral hemispheres, which are best developed in *D. inflatus,* are much smaller proportionately than in existing dogs and narrow forward more decidedly because of the undeveloped frontal lobes. The temporo-sphenoidal lobes, on the other hand, are relatively large and give the posterior part of the cerebrum its great vertical diameter. Apparently, the hemispheres leave the olfactory lobes entirely uncovered and extend but little over the lateral lobes of the cerebellum.

The sulci of the cerebral cortex, contrary to what might have been expected, are a little more complex than in the succeeding genus, *Cynodesmus,* of the lower Miocene.

As in the latter, the crucial fissure, so characteristic of existing Fissipedia, is not indicated at all; a longitudinal blood-vessel, with short side-branches, follows the line of meeting of the two hemispheres, but is not visible in *Cynodesmus.* As in the latter, the lateral and supra-sylvian sulci are straight and do not curve downward over the temporo-sphenoidal lobe, but there is an additional short sulcus on the occipital lobe, internal to the lateral fissure. No trace of this short posterior sulcus appears in *Cynodesmus,* but no great importance can be attached to such a minor detail, for the cast reproduces the inner surface of the skull, and but imperfectly represents the vanished brain. The crucial fissure is so widely distributed among the families of existing Carnivora that it must, almost certainly, have been inherited from a common ancestry and must, therefore, have appeared in the Eocene, yet no brain-cast of an Oligocene or Miocene carnivore known to us shows any sign of it. Very probably the absence of this characteristic sulcus is due rather to the failure of the brain-case to register it than to the lack of it in the living brain.

The skeleton of *Daphoenus* was very fully described by the late Mr. J. B. Hatcher from the fine specimen which he collected and which is now in the Carnegie Museum of Pitts-. burgh. As it is desirable that this monograph shall contain all the essential information concerning White River mammals, Hatcher's account of the limbs and feet is reproduced here in abbreviated form and after verification and correction by comparison with the original skeleton in Pittsburgh. Other relevant material in the New York and Princeton museums has also been made use of.

SKELETON OF THE TRUNK

The backbone has been recovered with considerable completeness, especially in the Pittsburgh skeleton, which lacks the neck and the sacrum. Even so, however, it is not yet possible to give the vertebral formula with entire confidence; according to Hatcher it was: C 7, D ? 13, L ? 7, S 3, Cd ? 23.

Cervical Vertebrae. No perfect example of the *atlas* is available for description, but incomplete ones afford important information; the vertebra is elongate antero-posteriorly and its inferior arch is very slender. As in the cats, the canal for the vertebral artery pierces the hinder border of the transverse process and has a horizontal direction, while in existing Canidae the canal opens in advance of the posterior border and pierces the transverse process almost vertically.   The posterior cotyles for the axis are small and nearly plane and are more distinctly separated from the articular surface for the odontoid process of the axis than in modern dogs, in which the three facets are confluent.   The neural arch is low and broad, considerably elongated from before backward and with only an inconspicuous tubercle for the spine.

The *axis* is likewise feline rather than canine in its general character and appearance. The centrum is elongate, narrow and depressed, with a thin and inconspicuous hypapophysial keel, running along the ventral surface, and has a slightly concave posterior face. The articular facets for the atlas are convex and rise higher upon the sides of the neural canal than in *Canis*, and on the ventral side they project below the level of the centrum, so that they are separated by a broad notch, which is not present in the modern dogs and is not well marked in the cats.   The odontoid process is a long, slender, bluntly pointed peg, with a heavy rounded ridge upon its dorsal surface, which is continued back along the floor of the neural canal.   The transverse processes are rather long and relatively very stout; they are shorter and heavier than in *Canis*, and keep more nearly parallel with the centrum. The vertebrarterial canal is longer than in the modern dogs and its posterior opening is not visible from the side.   The neural spine is the great, hatchet-like plate usual among the Carnivora, but is more cat-like than dog-like in shape.   In existing Canidae the spine is not prolonged behind the post-zygapophyses as a distinct process, but its hinder borders curve gently into them.   In *Daphoenus*, as in most cats and viverrines the spine extends behind the post-zygapophyses as a blunt and thickened process and is separated from them by a deep notch.

The other cervical vertebrae are more slender and lightly constructed than in existing dogs of corresponding stature.   The centra are long, narrow, depressed and feebly keeled and the keel does not terminate in a posterior hypapophysial tubercle, such as appears in *Canis*, except in the largest individuals, which have some indication of it; the transverse processes do not differ in any important way from those of the modern genus, but the neural arches are very different in being much narrower antero-posteriorly than in *Canis*, so that the interspaces between the arches of successive vertebrae are much larger and are longer antero-posteriorly than broad transversely.   The neural spines are likewise different from those of Recent dogs.   The third cervical has no spine; on the fourth there is a very low spine and on each successive vertebra the spine becomes higher and more pointed; that of the seventh is very high and slender, and much more prominent than in *Canis*; it is almost as high, though not nearly so stout as is the spine of the first dorsal in the latter.

Dorsal Vertebrae.   It is probable that the extinct genus agreed with its modern relatives in having twenty trunk-vertebrae, thirteen dorsal and seven lumbar.   Compared with the skull and the lumbars, the dorsals seem small and light and the thorax short.   The first dorsal has a broad, much depressed centrum, with convex anterior face and deeply concave posterior.   The pre-zygapophyses project forward strongly and, as in the cervicals, the notch between them is deeply incised, invading the base of the neural spine, a very different condition from that of *Canis*.   The post-zygapophyses are much smaller, but project prominently from the hinder end of the neural arch, both laterally and posteriorly.   The neural spine is high and compressed, shaped very much as in *Canis*, but rather more slender.   The transverse processes are long, heavy and prominent, bearing on the ends deeply concave pits for the tubercles of the first pair of ribs.

The second dorsal is very like the first, but has a smaller, narrower, lighter and much less depressed centrum; the pre-zygapophyses are smaller, and less concave and not so widely separated, while the post-zygapophyses are larger and face directly downward; the transverse processes are much smaller in every dimension and spring from the neural arch at a higher level, but the facets for the tubercles of the second pair of ribs are relatively large. The other vertebrae in the anterior part of the thoracic region have small centra and, in general, are very much like those of *Canis*. The ?sixth vertebra has a curiously shaped spine, which exaggerates the condition in the modern genus; the proximal portion is strongly inclined backward, while the distal portion is curved so as to project upward and the succeeding vertebrae, as far as the ?tenth, have similar spines.

The anticlinal vertebra, as in existing dogs, is probably the tenth and, at this point, the dorsals undergo an abrupt change, becoming like lumbars. In *Canis* the spine of the tenth dorsal is small and much lower than those of the ninth and eleventh, but in *Daphoenus* the spine is much better developed both in height and thickness; the post-zygapophyses are small, somewhat convex and placed high upon the neural arch, presenting outward. The ?twelfth and ?thirteenth vertebrae are much like lumbars, except for the smaller and lower spines, which are thickened at the distal end, and for the entire absence of transverse processes, which in *Canis* are present, though very short. The anapophyses are remarkably long and stout and are much heavier and more prominent than in the recent genera of the family. High, massive, metapophyses arise from the pre-zygapophyses.

The *lumbar vertebrae* presumably numbered seven, though not more than six have been found associated with any one skeleton. These vertebrae are remarkable for their relatively great size and massiveness and for the unusual development of all their processes; in these respects they are feline rather than canine in character and appearance. The following description assumes that there were normally seven lumbar vertebrae and that the partial skeleton described by the senior author in 1896 had lost the third. The centra increase in length posteriorly, reaching a maximum on the ?fifth and ?sixth, but the ?seventh is no longer than the first, though much broader and heavier. Compared with those of *Canis*, these centra are longer, stouter, less depressed and more rounded. The transverse processes are intermediate in character between those of *Canis* and of the larger species of *Felis*, they are relatively longer and heavier than in the former, less so than in the latter, which, however, they further resemble. In Recent dogs, the processes are directed horizontally outward and forward; in *Daphoenus*, as in the cats, they have a downward, outward and forward direction.   .

The neural spines are also intermediate in character between the Recent dogs and the larger felines; they are much higher, broader antero-posteriorly, more thickened at the distal end and more steeply inclined forward than in the former, and in the larger individuals particularly, the height of these spines is very striking and the likeness of the lumbars to those of the contemporary sabre-tooth *Dinictis* is remarkably close. Another similarity between the lumbars of *Daphoenus* and those of the cats is in the height and massiveness of the metapophyses, which are much better developed than in Recent dogs, though they are greatly reduced, almost vestigial on the last lumbar. The anapophyses are smaller than on the posterior dorsals and diminish in size on each successive vertebra; only on the first and second are they very large.   In existing Canidae, these processes are vestigial except on the first lumbar, where they are small.

The development of the anapophyses is another item of resemblance to the cats and emphasizes the statement, already made, that the posterior dorsal and lumbar vertebrae of this Oligocene dog are decidedly more feline than canine in appearance.   No doubt, the resemblance extended to the method of capturing prey.   Wolves and Dholes (the Indian Wild Dog) take their prey by running it down and, to that end, they are admirably adapted "cursorial machines," as Huxley said of the horse.   Cats, on the contrary, stalk their victims and then overpower them by a tremendous leap, or series of bounds.   The limbs and feet of *Daphoenus* do not suggest powers of high and sustained speed, but the loins are very suggestive of great leaping ability.

The *sacrum* is made up of three vertebrae and, in correlation with the great develop-ment of the tail, it resembles in many respects that of the larger cats; only the first sacral is in contact with the ilia and has massive pleurapophyses.   The centra of the sacrals are much larger and heavier than in modern dogs and the post-zygapophyses are much more prominent, but the resemblance of the sacrum of *Daphoenus* to that of the larger cats is not very close; the neural spines are lower and weaker; the neural canal is smaller and the transverse processes are decidedly shorter.   From the sacrum of existing dogs, that of the Oligocene genus differs especially in its greater proportionate length and massiveness.

The *caudal vertebrae* are still imperfectly known, for no complete series of tail-vertebrae has been recovered.   The Pittsburgh skeleton has the greatest number, fifteen, yet found in association and one of the Princeton specimens has thirteen, another eleven.   These suffice to show that the tail was remarkably long and thick and was, in fact, as well devel-oped as that of the Tiger, or the Leopard and, therefore, much longer and thicker than in any of the existing Canidae.

The first vertebra of the tail resembles that of the Lion, but is relatively lighter and more slender and has a short, but distinct neural spine; the transverse processes are very long, but not so broad as in the Lion, and are strongly recurved.   The ?fifth caudal has post-zygapophyses which extend beyond the end of the centrum; the hinder transverse processes point outward and backward; facets for the chevron-bones first appear on this vertebra.   The ?seventh caudal has well developed transverse processes, meta- and zyga-pophyses.   The ?ninth has prominent transverse processes, which are much more marked than in modern dogs, or even cats, and the prominences for attachment of the chevron-bone are very conspicuous.   In the middle part of the tail the vertebrae are extraordinarily elongate and resemble those of the long-tailed cats, but are relatively longer and more slender; the neural canal is absent from the ?thirteenth vertebra and the various processes are reduced to tubercles.   Near the end of the tail, the vertebrae become slender rods.

The anterior *ribs* are short and relatively very stout and all of them are shorter than in those modern dogs which have equally long limbs; the thorax was evidently proportionally less capacious than in the latter.

. The *sternum*, so far as can be judged from incomplete material, consisted, as in Recent dogs and cats, of the manubrium, xiphisternum and seven mesosternal segments, which resemble those of the dogs in general character, but are somewhat more slender.

The *baculum*, which articulates with no other bone and is connected with the pubes by ligaments, is but rarely found, being almost invariably lost in fossilization; happily it remained in natural association with the Carnegie Museum skeleton, for this element has much systematic and evolutionary interest.   Its occurrence among modern Fissipedia is seemingly capricious and without significance; in the aeluroid superfamily (in the restricted sense of the term) the *os penis* is either vestigial, or lacking, as is also true of the Herpestoidea, except the cat-like *Cryptoprocta*, in which it is large and well-developed.   In the dogs it is somewhat reduced, being straight, grooved on the dorsal side, and pointed, without distal enlargement, or perforation.   It is much the best developed in the raccoons and mustelines, in which it is large, especially near the proximal end, with a strong sigmoid curvature, enlarged and perforate, or furcate at the distal end.

In *Daphoenus* and in its contemporary canine *Pseudocynodictis*, the baculum is large, fully developed and generally raccoon-like, though there are considerable differences between the two Oligocene genera; it is much less decidedly curved and more dog-like in the former.   Of this bone in *Daphoenus*, Hatcher writes: "The os penis throughout the proximal two-thirds of its length is elliptical in cross-section with the greater diameter directed vertically.   Proximally, it is much compressed into a flattened, wedge-shaped, very rugose extremity, for muscular attachment to the pubes.   Distally, the bone becomes more cylindrical in cross-section, and at about the middle of its length, a shallow groove appears on its inferior surface.   This gradually becomes more pronounced, giving rise anteriorly to a deep channel, and at a distance of 10 mm. from the extremity the bone is entirely bisected and sends forward . . . two peculiar spout-like processes, . . . each with a shallow groove on its internal surface" (p. 82).

The occurrence of these elaborately formed bacula in both subdivisions of the Oligocene Canidae and in the modern *Cryptoprocta*, indicates that this was the original form of the bone in the suborder, which has been retained in the "arctoids" and *Cryptoprocta*, simplified and reduced in Recent dogs and lost almost completely in the cats, hyenas and viverrines. This seems to be a much more rational conclusion than to assume that in the Madagascar genus, the bone was retained, or independently developed and that the Oligocene dogs were in no way related to the modern Canidae, which from the beginning of their history, had a different type of baculum.

FORE LEG AND FOOT

Of the *scapula*, nothing is yet known.

The *humerus*, especially its distal portion, is decidedly more feline than canine in character, though, as will be subsequently shown, such similarities do not necessarily, or even probably, imply any close relation between *Daphoenus* and the Felidae.   The head is sub-elliptical in shape and the bicipital groove deep, the external tuberosity much exceeds the internal one in size.   The shaft is rather short and stout and is less arched than in *Canis;* the proximal part of the deltoid ridge, at least in the Carnegie Museum skeleton, is not very prominent, but it descends low upon the shaft and becomes very conspicuous distally, much more so than in existing canines and felines, but does not attain any such exaggerated development as in the White River sabre-tooth cats, such as *Drepanodon* and *Dinictis.*

The distal end of the humerus is remarkably cat-like in structure and suggests no relationship to the modern Canidae.   The supinator ridge, which in *Canis* is almost obsolete, is very prominent and extends far up on the shaft.   The internal epicondyle is very much larger, more rugose and prominent than in the modern genus, quite as much so, indeed, as in the cats and, as in the latter, there is a large entepicondylar foramen, bridged by a stout, straight bar of bone.   The anconeal fossa is lower, broader, shallower and altogether more cat-like than in *Canis* and does not perforate the shaft in a supra-trochlear foramen.   The humeral trochlea is extremely low, its vertical diameter being conspicuously less than in *Canis*, or even in *Felis*, resembling in this respect the sabre-tooth *Drepanodon*.   The shape of the trochlea is feline in character, having a simply convex surface for the capitellum of the radius and no such distinct intercondylar ridge, or convexity as occurs in Recent dogs; the internal border of the trochlea is prolonged downward as a sharp and prominent flange.

The *radius* is singularly cat-like in structure and in all its parts is much more feline than canine.   The proximal end is an oval capitellum for articulation with the humerus, its transverse diameter but slightly exceeding the antero-posterior dimension.   The anterior notch of the humeral surface is somewhat more deeply incised than in *Felis*, but not more than in *Drepanodon* (*Hoplophoneus*) which has an entirely similar capitellum.   The articular facet for the ulna surrounds the capitellum for somewhat more than half of its circumference, which is in remarkable contrast to the small size of this surface in *Canis*.   The shapes and modes of articulation of the bones which make up the elbow-joint in *Daphoenus* show that these animals retained unimpaired the power of pronation and supination of the fore-paw. In existing Canidae, on the other hand, this power is lost; the head of the radius is so broad as to occupy nearly the entire width of the humeral trochlea and so interlocks with it as to permit only the movements of flexion and extension.

The shaft of the radius is short and stout and in shape resembles that of the cats and is thus very different from the broad, antero-posteriorly compressed and almost uniform radial shaft of the Recent Canidae.   The bicipital tubercle is prominent and between it and the capitellum the shaft is distinctly constricted.   The distal portion of the radius is also very feline, but relatively lighter and narrower.   In the Pittsburgh skeleton there is on the inner side of the distal end of each radius a large and conspicuous exostosis, which has a very pathological appearance and, yet, its exact duplication and symmetry in both extremities seem opposed to this view.   The same bony growth has been observed in other individuals, but not in all, and its real nature is difficult to understand.   On the dorsal side of the distal end are well-marked sulci for the extensor tendons of the digits.   The distal facet for the ulna is small, subcircular in outline and forms a distinct prominence.   The carpal surface has a shape like that in the cats and is more concave transversely and narrower palmo-dorsally than in Recent dogs, and its inner border is more extended into a distally projecting flange.

The *ulna* is hardly less characteristically feline than the radius; the olecranon is rather short and its antero-posterior dimension is relatively less than in either *Felis* or *Canis* and its postero-superior end is thickened and rugose, but somewhat less so than in the modern genera named; the tendinal sulcus is wider and deeper than in *Canis*, less so than in *Felis*. The sigmoid notch is deeply incised and the coronoid process very prominent; the proximal humeral facet is relatively much wider than in *Canis* and is continuous with the broad infero-internal facet, while the infero-external facet is almost obsolete.   The radial surface is broad, deeply concave and single, while in *Canis*, though much smaller, it is divided by a sulcus into two portions.   The shaft of the ulna is stout; its proximal portion is laterally compressed, but it tapers toward the distal end, becoming trihedral in section.   This shaft is very feline and differs entirely from that of Recent Canidae, which has become very much more reduced and styliform, a change which is obviously correlated with the increased size of the radius.   The distal end is narrow and carries a convex, undivided surface for the pyramidal and pisiform.   The distal facet for the radius is placed on a prominent projection.

Hatcher made some very interesting observations upon the forearm bones of *Daphoenus*.   In substance he says: The ulna and radius are remarkably short, not only in proportion to the humerus, but also in comparison with the modern Canidae and Felidae. In the Carnegie Museum skeleton, the humerus has a length of 185 mm. and the radius one of 135 mm., a ratio of approximately 12 to 9.   In the Tiger the humerus is 270 mm. and the radius 240 mm. long, a ratio of 9 to 8; while in the Coyote (*Canis latrans*) the radius is longer than the humerus, the radius measuring 163 mm. and the humerus 157 mm.   These bones are not only much shorter proportionately than in *Canis*, but they are also stouter and more completely crossed, both of which characteristics are feline rather than canine.

In the *carpus*, the scaphoid, lunar and central are ankylosed to form the *scapho-lunar*, which is much the largest of the carpal bones; it differs markedly from that of Recent dogs and cats and is more like that of the White River sabre-tooth *Drepanodon*.   It is broad transversely, thick palmo-dorsally, but very short proximo-distally; the tubercle at the postero-internal angle of the bone is distinct though smaller than in the existing dogs or cats.   The radial facet is simply convex in both directions, not having the postero-internal saddle-shaped extension which occurs in *Canis;* this radial surface is reflected far over upon the dorsal and internal sides of the bone.   Distally, there are four facets for articulation for the four carpals of the second row.   There is no lateral contact with the pyramidal, as there is in *Canis*, being separated from it by the narrow proximal edge of the unciform.

The *pyramidal* is much flattened, more so than in the cats or the modern dogs; on the palmar side it gives off a considerable process, which overlaps and supports the head of the fifth metacarpal.   On the distal end the pyramidal has a large concave surface for the unciform and on the proximo-external side are separate articular surfaces for the ulna and pisiform respectively.   The *pisiform* is peculiarly shaped, having a narrow neck and very broad tuberosity, broader than in *Canis*.

The proportions of the *trapezium* and *trapezoid* are the reverse of those in *Canis*, the former being the larger and having a large distal facet for mc. I, but hardly touching mc. II. The *trapezoid* is a small, triangular bone, with apex toward the palmar side; it slightly overlaps the magnum, as in the cats, and of the metacarpals is in contact with only mc. II. The bone is more feline than canine in character.

The *magnum* is very short proximo-distally, but on the ulnar side it rises into a sharp keel between the pyramidal and scapho-lunar, a feline character; its principal metacarpal contact is with mc. III, but it has also a very limited contact on the radial side with mc. II and on the ulnar side with mc. IV.

The *unciform* is, after the scapho-lunar, the largest of the carpals; in form it is intermediate between that of *Felis* and of *Canis*, being wedge-shaped, with the apex proximal, which separates the pyramidal from the scapho-lunar and makes the contact of both of those carpals with the unciform lateral rather than distal.   As usual, the unciform carries mc. IV and V.

As a whole, the carpus is remarkably broad transversely and short proximo-distally and very feline in character.

The *metacarpus* is likewise very short and broad and far less differentiated and specialized than in the existing Canidae. In order to make clear the difference, in structure of the manus, between the Oligocene and Recent members of the family, it will be of advantage to preface the description with Schlosser's account of the characteristic features of the latter: "The metapodials are strikingly elongate. They display an almost square cross-section, in consequence of the mutual pressure; they are uncommonly closely applied to one another. . . . The distal articular surfaces have the appearance of very short cylinders. . . . The arrangement of the carpals is apparently more primitive than in the other Carnivora, at least as the carpals articulate with one another and with the metacarpals serially, instead of alternatingly. Only the scapho-lunar has attained considerable size; magnum, trapezium and trapezoid remain very short and all, proximally, as well as distally, lie in one plane. In consequence of this the proximal facets of the metacarpals are approximately in a single plane." It would be more accurate to say that the distal ends of the carpals form a single convex surface, which fits into the corresponding concave surface formed by the conjoined proximal ends of the metacarpals.

To this excellent diagnosis may·be added that the metapodials of both manus and pes are arranged in two symmetrical pairs, a median and longer pair, composed of the III and IV digits and a shorter, lateral pair, the II and V. The arrangement is characteristically artiodactyl and not so remotely so, as Schlosser seems to have thought. The whole structure of the limbs and; more particularly of the feet, is in adaptation to the cursorial habits of Recent Canidae, upon.which stress has repeatedly been laid and which serve to explain the differences between the Recent Canidae and all other Fissipedia, except the hyenas, which have somewhat similar feet.

The foregoing account of the metacarpals in modern dogs does not at all apply to *Daphoenus*, which plainly was not cursorial in habits. The distal facets of the carpus do not form a continuous curve, nor do the proximal facets of the metacarpals. The latter are not arranged in symmetrical pairs, but are each of different length and except at the proximal end, are not in close contact, but radiate from the carpus like the sticks of a fan. They are not elongate, but very short and, in cross-section are transversely oval, not square. The distal articular surfaces are hemispherical, as in modern fissipeds generally, not segments of cylinders. In short, the metacarpals are like those of unspecialized fissipeds generally and in their mode of articulation with each other and with the carpals are much as in *Gulo*.

The metacarpals are five in number, the pollex being relatively large and unreduced. The *first metacarpal* is actually little longer than in the Coyote, but much longer in proportion to the other metacarpals, as well as stouter and in every way better developed; the proximal end is thick and heavy and has a large facet for the trapezium. The shaft is short and arched toward the dorsal side; the distal trochlea is hemispherical and has a palmar carina and the lateral processes for ligamentous attachment are more prominent than in *Canis*.

The *second metacarpal* is much longer and stouter than the first, though short in comparison with the limb bones, short as these are themselves. The proximal end is rather narrow, but very thick palmo-dorsally; the facet for the trapezoid is of more uniform width· than in *Canis*, narrowing less toward the palmar side; the ulnar border rises more above the level of mc. III and has a more extensive contact with the magnum than in the modern genus, less than in *Felis* and very much as in *Drepanodon*. The conjoined facets for the magnum and mc. III form a broad, curved band on the ulnar side of the head. The shaft is short, weak and of oval cross-section and is arched toward the dorsal side. The distal end is made broad by the large, rugose processes for the attachment of the lateral ligaments, which are much more prominent than in *Canis*. The distal articular surface is hemispherical and is demarcated from the shaft by a depression, such as does not occur in the modern Canidae.

The *third metacarpal* is the longest of the series and, though not much exceeding the fourth in length, is much longer than the second. The shape of the proximal end is quite as in *Canis*, except for the greater dorso-palmar thickness; the facet for the magnum is narrow, but deep palmo-dorsally. There appear to be differences between various individuals in the carpal articulations of mc. III and IV. In the Pittsburgh skeleton mc. III is excluded from the unciform by mc. IV, which has a minute contact with the magnum, but in a Princeton specimen the more usual articulation of mc. III with the unciform is displayed. The shaft is somewhat more slender than that of mc. II and is of more quadrate shape, the dorsal and lateral faces forming distinct angles. The distal articular surface is like that of mc. II in all essentials.

The *fourth metacarpal* has a narrow, but palmo-dorsally thick head, which projects prominently behind the plane of the shaft, which is slender and nearly straight, though somewhat arched toward the dorsal side. In length mc. IV nearly equals III and exceeds all the others; the distal end has less prominent projections for the lateral ligaments than III.

The *fifth metacarpal* is, after the first, the shortest of them all; the proximal end has a concave surface for the unciform, around which it extends almost to a contact with the pyramidal. The short shaft broadens distally more than any of the other metacarpals, but the hemispherical trochlea is smaller.

The *Phalanges* are thus described by Hatcher: They "are intermediate in character between those of the dogs and cats. Those of the proximal row are somewhat arched as in the cats. The second series are nearly symmetrical as in the dogs, but the distal articular surfaces are less expanded laterally than in the dogs and are continued farther back, upon the superior surface, as in the cats, indicating that the terminals were to a certain extent at least retractile. The terminal phalanges are high and very much compressed, claws with rudimentary hoods. They are distinctly cat-like, rather than canine in character.

"Taken as a whole, the forelimb and foot of *Daphoenus* was [sic] comparatively short, the forearm and foot especially so. In general, its structure is decidedly feline rather than canine, and this applies alike to the bones of the brachium, the antebrachium and the manus, though there are a few canine characters, more especially in the structure of the manus and the character of the proximal and second series of phalanges" (pp. 87–8).

HIND LEG AND FOOT

The *pelvis* is entirely lacking in the Pittsburgh skeleton, but in the Princeton collection are several broken specimens, which so supplement one another, that the character of the *os innominatum* may be determined with the exception of the crest of the ilium, which is unfortunately missing in all the individuals.

So far as it is preserved, the pelvis is feline rather than canine, both in general appearance and in the details of structure. The peduncle of the ilium is wider and shorter than in *Canis*, narrower than in *Felis;* the anterior plate expands to its full width somewhat more abruptly than in the latter, but has the narrow form which is found in the cats and is not so broad at the free end as in the dogs.   The gluteal surface is not simply concave, as it is in the two Recent genera named, but is divided into two unequal fossae, by a prominent longitudinal ridge, such as occurs, though less conspicuously, in certain viverrines.   This feature is repeated in another White River genus of the family, the little *Pseudocynodictis* and is almost exactly the same in the contemporary sabre-tooth *Dinictis*.   The sacral surface is not nearly so far forward of the acetabulum as it is in *Canis*, and occupies a position more as in the cats.   The ischial border of the ilium is, for most of its length, nearly straight and parallel to the acetabular border, but descends more abruptly than in either Recent dogs or cats and follows a course more as in *Viverra*.   As in *Canis*, the acetabular border is more distinctly marked than in the true felines, and ends near the acetabulum in a long, roughened prominence, the anterior inferior spine.   The pubic border is very short and, hence, the iliac surface is not well defined.   The acetabulum has its borders somewhat more elevated than in the cats.

The ischium, which in existing Canidae is much shorter than the ilium, is very elongate and relatively even longer than in the felines.   The anterior part of the ischium is straight, rather slender and of obscurely trihedral cross-section; behind the acetabulum the dorsal border rises into a convexity; the spine of the ischium is terminated abruptly behind by the sciatic notch, which is as conspicuous as in the cats, while in *Canis* it is very faintly marked.   The hinder part of the ischium is expanded into a broad and massive plate, which is very rugose on the external surface.   This posterior portion is not so strongly everted and depressed as in the modern dogs and there is no such stout and prominent tuberosity, another resemblance to the cats.

The pubis is L-shaped and its anterior, descending limb is unusually long, broad and thin, much more so than in the cats or the modern dogs.   The obturator foramen is very large and oval, in shape and size agreeing much better with *Felis* than *Canis*.

The *femur* is long in proportion to the fore-limb bones, but not elongate in comparison with the size of the skeleton.   While not differing markedly from the thigh-bone of *Canis*, it yet has some resemblances to that of the felines.   The small hemispherical head is set upon a longer neck than in Recent dogs and has a smaller, deeper, more circular pit for the round ligament than in them.   As in *Canis*, the head projects more obliquely upward and less directly inward than in *Felis*.   The great trochanter is large and rugose, but has no such antero-posterior extension, does not rise so high and is not so pointed as in the existing Canidae and the digital fossa is smaller and shallower.   From the great trochanter a sharp and prominent ridge, the *linea aspera externa*, descends along the outer border of the shaft; the proximal part of this ridge forms a prominence which may represent the third trochanter. The lesser, or second trochanter, is more conical and more prominent than in *Canis* or *Felis*.

The shaft of the femur is long, slender and nearly straight, though slightly curved forward; it differs from that of the modern dogs in its lesser curvature and in broadening and thickening more gradually toward the distal end, and from that of the true cats in being more slender and cylindrical.   The rotular trochlea is rather narrower than in the latter, or even in *Dinictis*, but resembles that White River sabre-tooth in its shallowness, its lack of prominence and proximo-distal shortness.   A decided difference from both *Canis* and

Felis is that the borders of the trochlea hardly project at all in front of the plane of the shaft.

The femoral condyles are feline rather than canine in shape; they are small, and of nearly equal size, though the outer one is slightly the larger of the two, and project much less prominently behind the shaft than they do in *Canis*, and are also narrower and less widely separated than in the latter. As in so many features of the limb-bones, the whole distal end of the femur is more like that of *Dinictis;* however, the rotular groove is shorter and broader and the condyles are even less prominent.

The *patella* is very different from that of the Recent Canidae, in which this bone is small, narrow and thick and has more resemblance to that of *Dinictis*. It is broad, thin antero-posteriorly; the anterior surface is more rugose than in the sabre-tooth genus and the proximal end is more pointed, not so abruptly truncated. The facet for the rotular trochlea of the femur, in correlation with the shortness and shallowness of that groove, is but slightly convex transversely and concave proximo-distally.

The *tibia* is relatively slender and has more resemblance to that of *Dinictis* than of *Canis*. The bone is longer than the radius and is also proportionally longer with reference to the femur and humerus. The proximal end is both broad and thick and the facets for the femoral condyles are of nearly equal size, with a high spine between them. The cnemial crest is long, extending far down on the anterior side of the shaft, as in the cats, and therefore much longer than in the Recent dogs. The tuberosities extend beyond the shaft much more than in the latter and on the distal side of the external one is the large facet for the head of the fibula. The distal end is conspicuously different from that of *Canis* and is more as in *Dinictis*, and intermediate between the two; the astragalar facets are less deeply concave and the intercondylar ridge lower than in the former, while the facets are deeper and the ridge higher than in the latter. The large transverse sulcus, which, in the Recent dogs, invades the astragalar surface does not appear in *Daphoenus*. The internal malleolus is very large and resembles that of *Dinictis*, save that the posterior border is more inclined and the distal end of the process narrower. The sulcus for the tendon of the *tibialis posticus* muscle is very distinctly marked, more so than in *Canis*. The distal fibular facet is large and like that of *Dinictis* and, consequently, much larger than in the Recent Canidae.

The *fibula* is much more *Dinictis*-like than it is canine in character, for, in Recent dogs, the fibula is greatly reduced and very slender and, for the distal half of its length, is in close contact with the tibia, whereas, in *Daphoenus* the two bones are in contact only at the proximal and distal ends; elsewhere the interosseous space is wide. The proximal end is relatively much heavier than in *Canis*, especially in the antero-posterior dimension, in which it considerably exceeds that of *Dinictis*, though the excess is chiefly due to a large tubercle, which projects from the hinder border and is much larger than in the sabre-tooth genus. The shaft of the fibula is slender, though stouter both actually and relatively than in *Canis;* it is laterally compressed and, for the most part, of oval cross-section, but size and shape vary irregularly from point to point. The distal end resembles that of *Dinictis*, though it is somewhat smaller, and forms a very stout external malleolus, at the postero-external angle of which is a deep sulcus for the peroneal tendons. The distal tibial facet is rather larger than in *Dinictis* and that for the astragalus somewhat smaller, the two together forming a high, narrow band. There is no contact between fibula and calcaneum.

The *pes* is of even greater interest than the manus and, by way of introduction, it will be useful to quote Schlosser's diagnosis of the hind-foot among the Recent Canidae: "Of course, the arrangement of the tarsals and metatarsals deviates less from that of other Carnivora than does that of the carpals and metacarpals, yet even here we find some not unessential modifications. The navicular is considerably reduced in width, so that it can no longer cover the external side of the astragalar distal end. Metatarsal II, which in other carnivores, is in contact with mt. III at only two points, here touches for its whole dorso-plantar diameter. In consequence of the shortening of the tarsus, the ascending process of mt. V has become very short. The phalanges have, like the metapodials, almost square cross-sections; the claws are very sharp, though but little curved, and have considerable length. The dogs are the most pronouncedly digitigrade of all Carnivora."

The *astragalus* is decidedly different from that of both *Dinictis* and *Canis*, though Hatcher regarded it as "feline throughout." The trochlea is low and but moderately grooved, decidedly more than in the sabre-tooth, less than in modern dogs, and the articular surface is not extended so far upon the neck as in the latter. The pulley is asymmetrical, the outer condyle considerably exceeding the inner in size. The neck is relatively longer than in *Drepanodon, Dinictis*, or even than in *Canis* and is directed more toward the tibial side of the foot; the head is depressed and strongly convex. The external calcaneal facet is hardly so large, or so oblique in position as in *Dinictis*, but is yet more as in the latter than in *Canis*. The sustentacular facet is shorter and wider than in Recent dogs, and the sulcus separating it from the external facet is very much shallower. The distal accessory facet for the calcaneum, which is so distinctly shown in *Canis*, is not present. On the head there is a small facet for the cuboid.

The *calcaneum* is more as in *Dinictis* than in the modern dogs, though the tuber is longer and more compressed than in either of these groups, and its dorso-plantar diameter is more uniform, increasing less toward the distal end; the free end is less thickened and more deeply grooved for the Achilles tendon. Along the fibular edge of the dorsal border is a deep and conspicuous groove, which occurs also in *Dinictis*, but not in *Canis*. The external astragalar facet is very like that of the sabre-tooth genus, being more angulated and more oblique in position than in modern dogs, presenting inward as much as dorsally. The sustentaculum also resembles that of *Dinictis* in being less oblique and much more prominent and in having its facet more widely separated from the external astragalar surface than in *Canis*. The articular surface is more regularly oval and more deeply concave than in the latter. In the modern Canidae there is a facet for the navicular, which forms a right angle with the accessory astragalar facet, just mentioned, but is not present in *Daphoenus* or *Dinictis*. As in the latter, there is a prominent projection from the fibular side of the calcaneum, near the distal end, which is not present in *Canis*, but occurs in many of the Recent viverrines, mustelines and raccoons.

The *cuboid* is not especially peculiar; it is relatively longer proximo-distally, narrower and thinner planto-dorsally than in *Dinictis*. The long, thick and rugose ridge on the fibular side overhangs the sulcus for the peroneal tendons and is more prominent than in the sabre-tooth, but lacks the great, rugose, plantar protuberance which is conspicuous in the existing Canidae, and the surface for the calcaneum is very much more convex than in the latter, in which it is nearly plane. The distal surface for the head of mt. IV is more concave than in *Canis*, while that for mt. V is smaller and lateral rather than distal.

The *navicular*, as compared with that of existing dogs, is short proximo-distally, but broad transversely, not having undergone the reduction in width which Schlosser describes as characteristic of the modern members of the family. The surface for the head of the astragalus is not more concave than in the latter and there is no such stout tubercle on the plantar side of the bone as occurs in them. Two very small facets on the fibular side, one near the dorsal, the other near the plantar border, articulate with the cuboid. The distal surfaces for the three cuneiforms have nearly the same shape and relative size as in *Canis*, but they are more in the same transverse line, that for the entocuneiform being less displaced toward the plantar side.

The *entocuneiform* is of similar shape to that of *Canis*, but larger, for this tarsal is functional and carries a fully developed hallux. The *mesocuneiform* is a very small, wedge-shaped bone, broadest dorsally and thinning to an edge on the plantar side. The navicular facet is concave and very different from the curious oblique surface in *Dinictis*. As is almost universal in the Carnivora, the proximo-distal diameter of this bone is much less than that of either of the two adjoining cuneiforms, which makes the head of the second metatarsal rise higher proximally than that of the first and third. The *ectocuneiform* is, as usual, by far the largest of the three, though it is relatively smaller than in *Dinictis;* in shape, it resembles that of *Canis*, with minor differences. Thus, the proximal end is less extended planto-dorsally and the navicular facet is more concave; the plantar tubercle has a more constricted neck and enlarged, rugose head; the facets on the tibial side for the mesocuneiform and second metatarsal are more distinct, while the distal facet for mt. III is more concave and has a shorter plantar extension.

As a whole, the character of the tarsus is rather more machairodont, or viverrine, than canine. A conspicuous difference from the tarsus of existing Canidae is that the articulations, which in the latter are nearly plane (*e.g.,* the calcaneo-cuboid), in *Daphoenus* retain the more primitive concavo-convexity.

The *metatarsus*, like the metacarpus, has five fully developed members, which are longer and most of them are more slender than those of the forefoot. The *first metatarsal* is considerably longer and stouter than the corresponding metacarpal; the head is enlarged both transversely and planto-dorsally and bears a rugose tubercle on the plantar side. The facet for the entocuneiform is large and strongly convex planto-dorsally, nearly plane transversely; no other articular surfaces are visible on the proximal end. The shaft is slender and arched dorsally, oval in cross-section, expanding at the distal end, where the breadth is increased by the prominent tubercles for the lateral ligaments. The distal trochlea is small, but well-formed and of irregularly hemispherical shape, with plantar carina. The first metatarsal of *Dinictis* is very like that of *Daphoenus* and certain viverrines, such as *Cynogale*, also have a hallux of similar proportions, but in all Recent Canidae, save certain domestic breeds, mt. I is reduced to a vestigial nodule of bone.

The *second metatarsal* is much longer than the first, but is relatively far shorter and weaker than that of *Canis* and rather resembles that of the viverrine *Cynogale*, but does not have the peculiarly shaped proximal end which characterizes that genus. In *Dinictis* mt. II is somewhat heavier than in *Daphoenus*, but is otherwise similar. In the latter the proximal end of mt. II rises considerably above those of mt. I and III, owing to the proximo-distal shortness of the mesocuneiform, and is, as it were, wedged in between the ento- and ectocuneiforms, an arrangement common to all the fissiped families and general among the creodonts.   On the fibular side is a wedge-shaped projection, which fits into a corresponding depression on the tibial side of mt. III, thus forming a very firm connection between the two bones; proximal to this projection are two facets for the ectocuneiform.   The shaft is straighter than in *Canis*, the distal end not curving toward the tibial side, as it does in the modern genus.   In cross-section, the shaft is transversely oval, while in Recent dogs, it has become trihedral for most of its length, owing to its close approximation to the shaft of mt. III.   The distal trochlea resembles that of *Dinictis* and differs from that of *Canis* in being more hemispherical and less semi-cylindrical in shape and in its demarcation from the shaft by a deep depression; the lateral ligamentous processes are likewise more symmetrical.

The *third metatarsal* is much longer and stouter than the second, the difference between the two being greater than in *Dinictis*, or in the viverrines, or even than in *Canis*.   The facet for the ectocuneiform is of the usual shape, but its plantar prolongation is shorter and broader than in existing dogs, and it resembles that of *Dinictis* in being oblique to the long axis of the bone, inclining decidedly toward the tibial side.   The tibial side of this facet is deeply incised to receive the wedge-shaped prominence of mt. II, an incision which does not appear in *Canis*, but is found, though rather less conspicuously, in *Dinictis*.   On the fibular side are two facets for mt. IV; one, near the dorsal border is a deep, hemispherical pit and the other, a small, plane surface, is on the plantar prolongation.   The shaft is nearly straight, except for a slight arching dorsally.

The *fourth metatarsal* forms a symmetrical pair with the third, very much as in Recent dogs and cats, though in *Daphoenus* these bones are relatively shorter and weaker.   In *Canis* mt. III and IV are closely appressed for most of their length and, in consequence of this, the shafts have taken on a more or less quadrate shape, with the approximate surfaces flattened, while the distal ends curve away from each other, somewhat as in the White River camel *Poëbrotherium*.   In *Daphoenus* only the proximal portions of the metatarsals are in contact and they diverge distally, though the radiation is somewhat less pronounced than is that of the metacarpals and, thus, the primitive oval cross-section is preserved. The two metatarsals are closely interlocked and in much the same manner as in *Canis*.   On the head of mt. IV are two facets for mt. III, of which the dorsal one is a stout hemispherical prominence, which is received into the pit on the head of mt. III already described.   The plantar facet is actually on the plantar rather than the tibial side; the prolongation from the head of mt. III extends around and embraces this facet and, by means of this double articulation, a very firm interlocking of the two bones is effected.   On the fibular side of mt. IV is a large and deep depression for a projection from the head of mt. V.   The facet for the latter is large, slightly concave and continues uninterruptedly from the dorsal to the plantar border, while in *Canis* there are two distinct and well-separated facets; the shaft resembles that of mt. III, but is more slender.

The *fifth metatarsal* forms a lateral pair with mt. II, which it almost exactly equals in length.   On the fibular side of the head is a very prominent projection, which ends in a rugose thickening and is directed obliquely outward and proximally, the "ascending process" of which Schlosser speaks in the passage above quoted.   In the Recent dogs this process is greatly reduced and in *Dinictis* it is large, but differently shaped, being a long and prominent ridge which projects proximally much more than externally, while in *Daphoenus* it is a blunt hook, extending outward much more than proximally.   The facet for the cuboid differs from that of *Canis* in being concave transversely and in facing as much to the tibial side as it does proximally, while in *Canis* this surface is plane and entirely proximal in position. On the tibial side is a rounded protuberance, which fits into the depression on the head of mt. IV; this protuberance is more prominent than in *Canis* and decidedly more so than in *Dinictis*. The shaft has a transversely oval cross-section, with sharp fibular edge and is thus different from the trihedral section, with flattened tibial side, which occurs in *Canis* and is much more like mt. V in *Dinictis*.

The *phalanges* exhibit a curious combination of characters; they are long both actually and proportionally, especially in comparison with the metatarsals. A *proximal phalanx* of one of the median digits is long and depressed, but strongly arched toward the dorsal side. The metatarsal facet has a different shape from that seen in modern dogs, the width being relatively greater and the dorso-plantar thickness less, and is more oblique to the long axis of the phalanx, presenting more dorsally and less completely proximally, and the notch for the metatarsal carina is less deeply incised. The distal trochlea, which in *Canis* describes a semicircle from the dorsal to the plantar surface, is much more restricted, projecting less prominently on the plantar side and not reflected so far over upon the dorsal face. On the other hand, this trochlea is more deeply cleft in the median line than in the existing genus and the tubercles for the attachment of the phalangeal ligaments are larger.

The differences from the modern Canidae, which have been enumerated are so many resemblances to *Dinictis*, in which the corresponding phalanx is somewhat shorter and wider than that of *Daphoenus* and has rather more prominent ligamentous tubercles, but, otherwise, is very like it.

The proximal phalanges of the lateral digits differ from those of the median pair only in being shorter, more slender and less symmetrical and in having a lateral curvature which becomes very pronounced in the hallux. In digit II the proximal end of the first phalanx is remarkably broad.

The *second phalanx* is of about the same length, relatively to the first as in *Canis*, but is broader, more depressed and more asymmetrical than in that genus. The proximal facet is more distinctly divided into two depressions by a higher median ridge and the beak-like process of the median dorsal border is much more prominent. The distal trochlea is reflected farther upon the dorsal side and is more prominent on that side, but less so on the plantar; it is thus more convex planto-dorsally, but much less concave transversely than in the modern genus. The asymmetry of this phalanx is marked; and the dorsal surface is hollowed near the distal end, allowing a limited retraction of the claws, which was accomplished in a different manner from this action in the cats. In felines and sabre-tooths the distal end of the second phalanx is so asymmetrically shaped as to allow the ungual to rotate past the shaft of the second and thus the retractility is complete. To find, in a dog, even a partial retraction of the claws is a very unexpected discovery, the significance of which must be carefully considered.

The *ungual phalanx* is short, laterally compressed and bluntly pointed; it is very little decurved and has a plainly marked groove on the plantar side near the distal end. These thin straight claws are in decided contrast to the heavy, conical and strongly decurved unguals of existing Canidae, though, among the latter, there is considerable variation in these respects. The articular surface for the second phalanx is much more deeply concave than in *Canis*, permitting a greater freedom of motion in this joint, as was necessary to effect the partial retraction of the claw.   The subungual process is not so large as in *Canis*, nor does it project so prominently from the plantar surface, but it is produced much farther proximally, as is generally the case in retractile claws.   The imperfect bony hood which envelopes the base of the horny claw is of about the same shape and size as in *Canis*, though the space between the hood and the body of the phalanx is narrower.

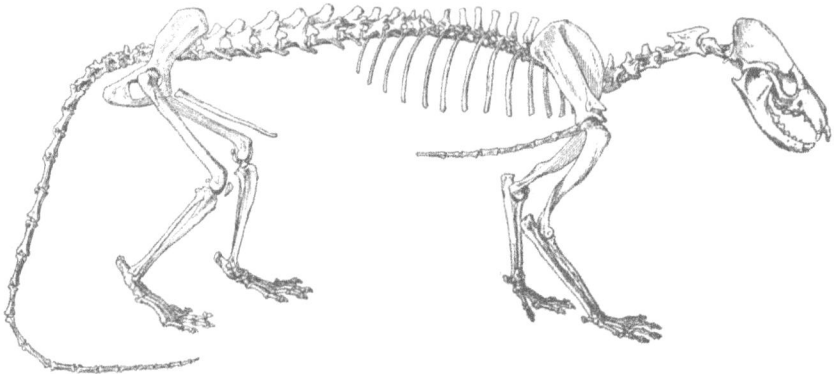

Fig. 5.   *Daphoenus vetus*, skeleton × 1/8, re-drawn from Hatcher, Carnegie Museum, Pittsburgh.

The Skeleton.   "The general aspect of the articulated skeleton is that of a long, slender-bodied, long-tailed and proportionately short-limbed carnivore.   In form and general proportions, the appearance of the skeleton is that of a cat, with a skull elongated as in the dogs.   The limbs are short in proportion to the length of the skull and vertebral column. The lumbar region is especially long and the lumbars exceptionally heavy.   The proportion of the axial to the appendicular skeleton is somewhat intermediate between that which obtains in the cats and creodonts" (Hatcher, p. 95).

This excellent summary gives an accurate conception of the appearance of the skeleton in these ancient and primitive dogs.   It should be further emphasized, however, that many, if not all, of the cat-like features are those which the later and even the Recent felines have retained from a much less specialized ancestry and which Recent dogs, with their high degree of cursorial specialization, have lost.   *Daphoenus* suggests a likeness in general appearance and proportions not only to the cats, but to many mustelines and viverrines as well and, except for the long, heavy, cat-like tail, to raccoons also.

Systematic Position of Daphoenus

This very difficult problem is full of paradoxes and seeming contradictions; only seeming, of course, for fuller knowledge will assuredly remove them.

As has been repeatedly emphasized in the foregoing pages, the dogs are pre-eminently runners and differ from all other Fissipedia (the hyenas are a partial exception) in the adaptation of their skeletons to high and sustained speed.   Other Carnivora stalk their prey by stealthy approach, or lie in wait for it on the branch of a tree and then, when the victim has come within reach, seize it by leaping upon it.   The dogs, and especially the foxes, also take prey in this manner, but their habitual method is by running it down through superior endurance and speed.   To this end the thorax is enlarged, so as to increase the lung-capacity, and the tail has lost much of its length.   The limbs and feet of all the later Canidae are greatly modified in a fashion that not remotely imitates the artiodactyl plan of structure.   Legs and feet are elongated, the ulna and fibula are much reduced, while radius and tibia are enlarged, the former so interlocking with the humerus, as to lose all power of rotation.   The feet are effectively tetradactyl, the hallux being completely lost and the pollex reduced to a vestige.   Of the four functional metapodials, the median ones, III and IV, form a symmetrical longer pair, and the laterals, II and V, a symmetrical shorter pair.   Of all Fissipedia the dogs are, as Schlosser remarked, the most completely digitigrade.

Skull and jaws underwent no such transforming adaptations as did the extremities, the principal change being in the great development of the brain, especially in the enlargement of the cerebral hemispheres, in consequence of which the sagittal and occipital crests lost much of their height and prominence and the cranial vault increased in capacity. The auditory bullae became larger, reaching to the paroccipital process and were firmly ankylosed in the base of the cranium.   Leaving aside *Otocyon*, nearly all existing Canidae have retained the primitive number of teeth except for the loss of the third upper molar, giving the formula: $m\frac{2}{3}$, while in *Daphoenus* it is $m\frac{3}{3}$.   In *Cyon*, of India, and *Icticyon*, of Brazil, the molars are still further reduced to $m\frac{1}{2}$, but existing dogs have the long jaws and relatively primitive type of tooth-structure which are displayed in the Oligocene genera.

At the present day, there is considerable variety among the Canidae, wolves, jackals, dholes, foxes, fennecs, "raccoon-dogs," bush-dogs, etc., yet, such is their uniformity of structure that, until comparatively lately, naturalists included them all in the genus *Canis*, excepting only the problematical South African *Otocyon*.   In the Miocene and Pliocene of the northern hemisphere, on the other hand, there were many "phyla," or genetic series of dogs, some of which reached enormous size, all of which are now extinct.   What gives particular interest and significance to *Daphoenus* is that in that Oligocene genus we have approximately the common ancestor of three, at least, of those phyla of the Canidae.   One of these series constitutes the main line of canine descent, by the ramification of which arose most of the genera now living.   It is not yet possible to follow the evolution of all these forms, but, in general, it seems very probable that the smaller species, *Daphoenus hartshornianus*, is very nearly what the common ancestor must have been.   Hatcher proposed to remove this species to a new genus which he called *Protemnocyon*, and there is much to be said for this course.   On the whole, however, we regard it as better to retain the earlier arrangement.

A succession of genera carried on this line through the upper Oligocene, *Mesocyon*, of the John Day (upper Oligocene) and Gering (lower Miocene) *Cynodesmus*, of the Harrison (lower Miocene), *Tomarctus* of the upper Miocene and lower Pliocene, if not themselves the actual evolutionary steps, indicate very closely what those steps must have been, and from the Pliocene *Tomarctus* most of the modern genera may be derived.

A second phylum is exemplified today by the Dhole, or Wild Dog, of India (*Cyon*), and the Bush Dog of Brazil (*Icticyon*) which in spite of their wide geographical separation, agree with each other and differ from all other canines in certain features of the dentition.   It is probable that most of the development of these genera took place in South America and

Asia respectively from migrants from North America, for there is a long hiatus in the latter continent. The John Day genus *Temnocyon* stands near to the common ancestor of the Asiatic and South American genera and, as Hatcher suggested, *Temnocyon*, in turn, would seem to have been derived from the White River *Protemnocyon inflatus*, which is nearly allied to *D. hartshornianus*.

Still a third phylum, that of the "Bear-like Dogs," comprising animals of extraordinary structural characteristics and of remarkable interest, has long been extinct. This series very probably took its rise in the larger species of *Daphoenus*, *D. vetus*, and may be traced through the upper Oligocene, Miocene and Pliocene, the successive stages increasing in size until in *Amphicyon*, *Hemicyon* and *Dinocyon*, of the upper Miocene and lower Pliocene, gigantic dogs, the largest in the history of the family, made their appearance and extended their range to the Old World; indeed, they were first discovered in France. The "bear-likeness" of these astonishing creatures consisted only in the enlargement of the tubercular molars and so ursine are these teeth that *Amphicyon* and its allies were long regarded as ancestral bears, but this view has, perforce, been abandoned on the discovery of more or less complete skeletons and the finding of the actual primaeval bear ancestors.

As to the many feline characteristics which have attracted the attention of all students of this group, a careful analysis leads to the conclusion that these likenesses are not indicative of a close relationship of the dog and cat families, except as all the Fissipedia are related to one another, for they were obviously all derived from the same ancestral stock. The feline characteristics of *Daphoenus* are, for the most part, such primitive features as the cats still retain, yet were originally common to all, or most of the fissiped families. Other resemblances seem to have been independently acquired by the two families in adaptation to similar habits of life. There is much evidence in support of Schlosser's view, that the cats are the most isolated of all the fissiped families and the farthest removed from the others.

Species. As is true of most of the White River mammalian genera, the number of species of *Daphoenus* which are entitled to recognition, offers an unsolved problem, further complicated by the uncertainty as to the number of alleged species which were actually living at the same time and place. The analogy of existing genera shows that but few, or only one, species of a genus can exist in a given area. Each species has its own range and, while there is an overlapping of ranges at the borders, it is quite impossible that so many species as have been named and described for most of the White River genera can have lived together. Though allowances must be made for the probable mingling of the bones of species which had very different habitats, upland and valley, forest, swamp and grasslands, by water transportation, yet too little attention has been paid to the differences due to sex and age and to individual, fluctuating variation. Save in exceptional circumstances, the modern naturalist will not found a new species without a large suite of specimens, while most species of fossil mammals have been proposed for a single, more or less fragmentary individual. It cannot be doubted that the number of White River species currently recognized must eventually be greatly reduced.

Three species of *Daphoenus* would seem to be distinct, two of which, *D. vetus* Leidy and *D. hartshornianus* Cope, have been found only in the Brulé substage and the third, ?*D. dodgei* Scott, only in the Chadron. Hatcher has described an additional species, which he assigned to a new genus *Proamphicyon nebrascensis*, but the differences listed in his diagnosis are insufficient for generic distinction and *P. nebrascensis* is of doubtful specific standing. This arrangement would admit two species of this genus, occurring in the Brulé substage, *D. vetus* Leidy, *D. hartshornianus* Cope.   ?*D. dodgei* Scott belongs in a different category, as being more ancient than the others, and not contemporary with either of them.

Daphoenus vetus Leidy

(Pls. X, XI, Figs. 1–9, XII, Fig. 2)

Daphoenus vetus Leidy, Proc. Acad. Nat. Sci. Phila., 1853, p. 393.
Amphicyon vetus Leidy, *ibid.*, 1854, p. 157.
Daphoenus felinus Scott, Trans. Am. Phil. Soc., N.S., XIX, p. 361.
Proamphicyon nebrascensis Hatcher, Mem. Carnegie Mus. Pittsb., I, p. 68.

This is the larger species of the genus and, as here interpreted, includes a considerable range of difference in size and proportions.   In a large series of well preserved specimens, the dimensions intergrade from one to another and the influence of age and sex must also be taken into consideration.   The larger and more robust skulls are, not improbably, those of males, while the smaller and more slender ones may be female.   Leidy's type is one of the latter sort, and the type of *D. felinus*, as well as the Pittsburgh skeleton, described by Hatcher under that name, are among the largest individuals.   Aside from stature, this species is characterized by larger upper molars, especially m3, which is generally implanted by two roots, though sometimes by one.   This is the only species of which any considerable part of the skeleton is known and, thus, comparisons are not yet feasible; the descriptions are given under the head of the genus.

Horizon: Brulé.

MEASUREMENTS (*fide* Hatcher)

Upper cheek-teeth series, length	76.0 mm.	?Fifth caudal, length of centrum	35.0 mm.
Upper molar series, length	38.0	?Seventh caudal, length	40.0
Upper canine, ant.-post. diam.	12.0	?Ninth caudal, length	44.0
Upper canine, transv. diam.	9.0	?Thirteenth caudal, length	44.0
P4 ant.-post. diam.	16.0	Baculum, length	166.0
P4 transv. diam.	10.0	Baculum, depth at base	16.0
M1 ant.-post. diam.	11.0	Humerus, max. length	185.0
M1 transv. diam.	17.0	Humerus, distal width	41.0
Lower cheek-teeth series, length	83.0	Radius, length	135.0
Lower molar series, length	32.0	Ulna, length	171.0
M1, ant.-post. diam.	17.0	Ulna, length below coron. proc.	137.0
Skull, max. length	205.0	Carpus, width	34.0
Skull, width over zygomata	119.0	Carpus, max. length	12.0
Sagittal crest, length	95.0	Mc I, length	25.0
Sagittal crest, max. height	28.0	Mc II, length	37.0
Cranium, width at postorb. constr.	31.0	Mc III, length	47.0
Cranium, max. width	58.0	Mc IV, length	45.0
Cranium, width over postorb. proc.	43.0	Mc V, length	35.0
Mandible, length cond. to inc. alv.	152.0	Phalanges of digit I, length	29.0
Mandible, height of condyle above ang.	30.0	Phalanges of digit II, length	46.0
Mandible, height of coronoid	68.0	Phalanges of digit III, length	54.0
Mandible, depth below m1	26.0	Phalanges of digit IV, length	52.0
?First caudal, length	19.0	Phalanges of digit V, length	45.0
?First caudal, width over trans. proc.	61.0	Femur, max. length	201.0

Femur, width of dist. end............ 32.0 mm.
Tibia, width of prox. end............. 36.0
Tibia, length...................... 179.0
Fibula, length..................... 168.0
Mt. I, length...................... 35.0
Mt. II, length..................... 49.0
Mt. III, length.................... 58.0

Mt. IV, length..................... 61.0 mm.
Mt. V, length...................... 50.0
Phalanges, digit I, length........... 33.0
Phalanges, digit II, length.......... 47.0
Phalanges, digit III, length......... 56.0
Phalanges, digit V, length........... 47.0

Daphoenus hartshornianus (Cope)

(Pl. II, Fig. 6, XII, Figs. 1–1c)

Daphoenus vetus Leidy, in part, *loc. cit.*
Amphicyon vetus Leidy, in part, *loc. cit.*
Canis hartshornianus Cope, Ann. Rept. U. S. Geol. and Geogr. Surv. Terrs., 1873, p. 505.
Daphoenus hartshornianus Scott, Trans. Am. Phil. Soc., N.S., XIX, p. 361.

Much smaller than *D. vetus* and with relatively smaller tubercular molars; m3 implanted by a single root.   The measurements are from two specimens in the Princeton Museum.

Horizon: Brulé.

MEASUREMENTS

	No. 13,580	No. 12,635
Upper canine, ant.-post. diam. at base.............	10.0 mm.	10.0 mm.
Upper canine, transverse diam. at base............	7.0	7.0
Upper cheek-teeth series, length.................	62.0	62.0
Upper premolar series, length....................	43.0	42.0
Upper molar series, length......................	21.0	25.0
P1, length.....................................	5.0	6.0
P2, length.....................................	9.5	9.0
P3, length.....................................	10.5	10.0
P4, length.....................................	14.0	15.0
P4, width......................................	11.0	10.0
M1, length.....................................	12.0	12.0
M1, width......................................	15.0	16.0
M2, length.....................................	7.0	7.0
M2, width......................................	11.0	12.0
M3, length.....................................		4.0
M3, width......................................		6.0
Lower cheek-teeth series, length.................	66.0	71.0
Lower premolar series, length....................	39.0	38.0
Lower molar series, length......................	29.0	30.0
P1̄, length.....................................	4.0	
P2̄, length.....................................	9.0	
P3̄, length.....................................	10.0	9.5
P4̄, length.....................................	12.0	12.5
M1̄, length.....................................	15.0	15.0
M1̄, width......................................	12.0	12.0
M2̄, length.....................................	8.0	9.0
M3̄, length.....................................	4.5	5.5
Skull, median basal length.......................	170.0	
Skull, max. length, occip. crest to prmx. incl........	184.0	
Skull, width over zygomata......................	106.0	
Sagittal crest, length...........................	80.0	
Cranium, width at postorb. constr...............	22.0	

Cranium, max. width.............................	60.0	
Cranium, width over postorb. proc..............	36.0	40.0
Face, length prmx. to orb. rim..................		76.0
Face, width at p4..............................		55.0
Face, width at p1..............................	29.0	30.0
Mandible, length cond. to inc. alv...............	112.0	
Mandible, height of cond. above angle...........	29.0	
Mandible, depth below m1......................	20.0	21.0

?Daphoenus dodgei Scott

(Pl. XII, Figs. 3, 3a)

? *Daphoenus dodgei* Scott, Trans. Amer. Phil. Soc., N.S., XIX, p. 362.

Only mandibles of this species have as yet been found and its generic reference is, therefore, somewhat uncertain; it is, so far as known, confined to the Chadron substage (*Titanotherium* Beds). The lower dental series is relatively short and the premolars are shorter antero-posteriorly than in the Brulé species, but relatively thicker and stouter. The posterior cusp of the premolars, which in the other species is confined to p4, is, in ? *D. dodgei*, present on p2 and p3 as well. M1 is decidedly more primitive than in the succeeding species and suggests that, were more complete material available, it would be necessary to refer this species to another genus.

The lower sectorial, m1, has a low, heavy anterior triangle, of which the protoconid is more cylindrical, less flattened and blade-like than in the other species; the metaconid is much larger than in those species and, when viewed from above, seems nearly as large as the protoconid; the whole crown is reminiscent of the Eocene family Uintacyonidae. The molar series is relatively shorter than in *D. vetus* and m3 is implanted in the ascending ramus. The horizontal ramus of the mandible is relatively shorter and much deeper vertically and thicker transversely than in the other species.

The following measurements are taken from two individuals in the Princeton Museum; No. 11,422 is the type.

Horizon: Chadron.

MEASUREMENTS

	No. 11,422	No. 13,601
Lower cheek-teeth series, length..................	45.5 mm.	39.0 mm.
Lower p1, length...............................	3.0	
Lower p2, length...............................	6.0	7.0
Lower p2, width................................	4.0	
Lower p3, length...............................	8.0	9.0
Lower p3, width................................	5.0	6.0
Lower p4, length...............................	11.0	12.0
Lower p4, width................................	5.0	5.5
Lower m1, length...............................	14.0	15.0
Lower m1, width................................	7.0	7.0
Mandible, depth at m1..........................	23.0	26.0
Mandible, thickness at m1.......................	16.0	11.0

In No. 13,601 the teeth, though individually slightly larger than in the type, are more crowded together, which shortens the premolar series. The relative thinness of the jaw is due to crushing.

Protemnocyon Hatcher

Protemnocyon Hatcher, Mem. Carnegie Mus., I, p. 99, 1902.

The status of this genus is uncertain, for the type is a skull which, though in a fine state of preservation, is that of an old animal with much worn teeth. The third molar is greatly reduced and the upper one is sometimes absent. The type-skull agrees closely in size with that of *Daphoenus hartshornianus*, from which *Protemnocyon* differs significantly in the character of the upper teeth; the sectorial has a larger internal cusp, but smaller than in *D. vetus;* the tubercular molars (m$\underline{1}$ and $\underline{2}$) are wider transversely and the internal portion has a less fore-and-aft diameter, so that the crowns tend to assume a more triangular and less quadrate shape.

A very conspicuous difference of this genus from *Daphoenus* is the far larger brain-capacity of the cranium and, consequently, the lower sagittal and occipital crests, which gives the skull the appearance of an enlarged *Pseudocynodictis* rather than *Daphoenus*.

Protemnocyon inflatus Hatcher

Protemnocyon inflatus Hatcher, *loc. cit.*

This species is of nearly the same size as *Daphoenus hartshornianus*, but its much larger and more vaulted brain-case immediately distinguishes it. Hatcher's diagnosis is: "*Char. sp.* Temporal constriction anterior to union of superciliary ridges. Frontals broad, gently concave medially, but convex laterally, indicating the presence of well-developed frontal sinuses. Inferior margin of mandible nearly straight. M$\underline{3}$ much reduced in size or absent. P$\frac{1}{1}$, $\frac{2}{2}$, $\frac{3}{3}$, $\frac{}{4}$ large and with broad heels."

Horizon: lower Brulé.

The following dimensions are from Hatcher's paper.

MEASUREMENTS

Greatest length of skull	167.0 mm.	Breadth of upper sectorial	9.5 mm.
Greatest breadth of skull	85.0	Breadth of upper m1	15.5
Length of sagittal crest	68.0	Breadth of upper m2	11.0
Breadth of cran. at post-orb. constr.	23.5	Ant.-post. diam. of canine at base	8.5
Greatest breadth of cranium	46.0	Transverse diam. of canine at base	6.0
Greatest breadth of frontals	36.0	Length of lower cheek-teeth series	65.0
Length of palate	81.0	Length of lower premolar series	38.0
Length fr. incis. alv. bord. to ant. of		Length of lower sectorial	16.0
orbit	65.0	Length of lower p4	11.0
Greatest length of mandible	116.0	Breadth of lower p4	5.5
Depth of mand. at m$\overline{2}$	19.0	Ant.-post. diam. of lower canine	9.0
Depth of mand. at p$\overline{2}$	16.0	Transverse diam. of lower canine	5.0
Height of coronoid proc.	56.0	Width of ant. cotyles of atlas	32.0
Length of upper cheek-teeth series	58.0	Length of centr. of axis incl. odont.	44.0
Length of upper premolar series	43.0		
Length of upper sectorial	15.0		

Brachyrhynchocyon nom. nov. Loomis

Brachicyon Loomis, Amer. Journ. Sci., N.S., XXII, p. 101, 1931.

This is a remarkably short-faced dog, which is nearly related and probably ancestral to *Enhydrocyon* of the John Day. The upper premolars, except p$\underline{4}$, are much reduced in size, though retained in undiminished number. P1 is minute and vestigial, p2 is very small and simple, without posterior cusp; p3 is broken away on both sides of the type-skull and nothing more can be affirmed of it than its manifestly small size. The sectorial, p4, has a very oblique shear and bears a large internal cusp.

The first molar (m1) is large, especially in transverse width, and the second (m2) is very much smaller, not one-half so large; the third has been suppressed, giving the same dental formula as in *Canis*.

The lower premolars are very small, especially p1̄, which is vestigial; the others increase successively in size and have very small basal cusps. The sectorial (m1̄) has a low shearing blade and a trenchant heel, with single median ridge, though a trace of the inner cusp remains. This character is also displayed in a more perfect way by the John Day genera *Temnocyon* and *Enhydrocyon*. The second molar (m2̄) is small and the alveolus for m3̄ is so minute, that the tooth cannot have been more than a vestige.

Except for the relatively shortened face, the skull is very much like that of *Daphoenus hartshornianus*, which it closely approximates in size, and it has a similar small and loosely attached tympanic bulla. The coronoid process of the mandible has a more nearly vertical anterior border than in *Daphoenus* and the vestigial m3̄ is inserted in it.*

Brachyrhynchocyon intermedius nom. nov. Loomis

Brachicyon intermedius Loomis, *loc. cit.*, p. 101.

This, the only species of the genus yet made known, is of the size of *Daphoenus hartshornianus*, which resembles it, though generically distinct.

Horizon: Brulé.

MEASUREMENTS (*fide* Loomis)

Length of skull	149.0 mm.
Length of upper premolar-molar series	45.0
Length of lower premolar-molar series	53.0

Pseudocynodictis Schlosser

(Pls. XI, XIII)

Amphicyon Leidy (*nec.* Lartet), in part, Proc. Acad. Nat. Sci., Phila., 1853, p. 90.
Canis Cope (*nec.* Linn.), Synop. New Vert. fr. the Tert. of Colorado. Washington 1873, p. 9.
Galecynus Cope (*nec.* Owen), Bull. U. S. Geol. & Geogr. Surv. Terrs., VI, pp. 165–181.
Cynodictis Scott (*nec.* Bravard et Pomel), Trans. Amer. Phil. Soc., N.S., XIX, p. 364, 1898.
Nothocyon Wort. and Matt., Bull. Amer. Mus. Nat. Hist., XII, p. 124, 1899.
Pseudocynodictis Schlosser, Geol. Paläont. Abhandl., Jena (N.F.), Bd. V (XII), pp. 117–258.

Great confusion still prevails concerning the generic names which are properly applicable to the smaller canids of the North American Oligocene. For the last thirty years or more, the usual practice has been to refer the small White River dogs to the European genus named *Cynodictis* by Bravard and Pomel. Sometimes the *Nothocyon* of Wortman and Matthew has been made to include the White River species, but the authors named excluded these species from their new genus, the typical species of which is the existing South American *Canis urostictus* of Mivart. Only the small John Day species, which Cope had

---

* This term is preoccupied by *Brachycyon* Filhol, 1872; Professor Loomis has requested us to substitute *Brachyrhynchocyon* for it.

referred to the European *Galecynus*, were included in *Nothocyon*. Schlosser was the first to point out that the American Oligocene forms constituted a separate genus, for which he proposed the name *Pseudocynodictis*.

The most important distinction between the European and American genera is in the structure of the auditory bulla, which in *Cynodictis*, is small and either not ossified, or, more probably, so loosely attached to the skull, that it is always separated and lost in the fossil. In *Pseudocynodictis*, on the contrary, the bulla is large, inflated, heavily ossified and so firmly ankylosed to the base of the cranium, that it is almost invariably preserved in the fossil state. The present genus differs from *Daphoenus*, not only in its uniformly smaller size, and in the development of the tympanic bulla, but also in the marked reduction of the tubercular teeth; m$\underline{3}$ has been lost, while m$\underline{1}$ and $\underline{2}$, and $\overline{2}$ and $\overline{3}$ are very small; m$\overline{3}$ is especially minute. The formula is as in *Canis:* i$\frac{3}{3}$, c$\frac{1}{1}$, p$\frac{4}{4}$, m$\frac{2}{3}$.

DENTITION

All of the teeth are proportionately small, especially in *P. microdon*, which has much reduced premolars, but only m$\underline{3}$ has been lost. The premolars and sectorial molars have, when in a perfectly unworn state, high and very sharp-pointed crowns, but this height is speedily reduced by abrasion, the tips being worn away.

Upper Teeth. The minute *incisors* are very narrow and chisel-shaped and they increase slightly in size from the median to the lateral one, which is separated from the canine by a very short diastema; i$\underline{3}$ does not exceed i$\underline{1}$ in size nearly so much as it does in *Canis*, or *Daphoenus*. The *canine* is relatively smaller and thinner than it is in Recent dogs and has quite different proportions from those of *Daphoenus;* the root is stout and causes a rather conspicuous swelling of the maxillary.

The *premolars* increase in size by regular gradations from the first to the third, while the sectorial (p$\underline{4}$) is much larger. The premolars have simple, compressed and acutely pointed crowns; the posterior cuspule is very minute on p$\underline{1}$ and larger on each successive tooth, especially on p$\underline{3}$, where it almost forms a cutting blade. The sectorial, p$\underline{4}$, is small and has a more oblique position than in modern dogs; the internal cusp is greatly reduced and is much smaller than in *Daphoenus*, *Urocyon*, or even in *Canis*. The *molars* are small, especially the second (m$\underline{2}$); m$\underline{1}$ has a pattern very like that of *Daphoenus;* fundamentally it is tritubercular, with two external and one internal conical cusps and intermediate conules; the internal cingulum forms a crescentic shelf, which is much less developed than in the Recent genera, or even in *Daphoenus*. The second molar is very small, but similar to the first in pattern, except for the smaller size of the postero-external cusp and the lack of the posterior conule.

Lower Teeth. The *incisors* are very small and set close together, i$\overline{2}$ having its root behind the others. The *canine* is long, sharp-pointed and even more compressed than the upper one. The first *premolar*, which is separated from the canine by a very short space, is extremely small and is inserted by a single root; at the hinder end, the cingulum forms a minute basal cusp. P$\overline{2}$ is much larger than p$\overline{1}$ and is carried on two roots; the crown is of more regularly compressed-conical shape, with median apex and more distinct basal cusp. P$\overline{3}$ resembles p$\overline{2}$, but is larger, has a more prominent posterior cingular cusp and, in addition, a conule on the hinder border of the principal cusp, giving the tooth a triconodont appearance, in side-view.   Except for its larger size, better defined posterior cuspules, and anterior basal cuspule, p4̄ is like p3̄.

The *molars* are proportionately small; the sectorial, m1̄, has a high anterior triangle, especially the paraconid; the inner cusp is small, but acute and well-separated; the heel is basin-like, with external ridge higher than the internal one.   The tubercular molars are much reduced and evidently on the way to suppression; m2̄ has all the elements of m1̄, but they are very low and small, and m3̄, which is inserted by a single root, is tiny and can have had no functional importance; in the unworn state three minute points represent tubercles. In some individuals, this tooth is absent, without even an alveolus to indicate its presence in life.

SKULL

The skull of *Pseudocynodictis* is, in most respects, of the primitive canine type, yet resembles in appearance that of such viverrine genera as *Paradoxurus*.   As in the contemporary White River dog, *Daphoenus*, the facial, or preorbital region is very short in proportion to the elongate cranium.   The dorsal contour of the skull rises from the low occiput to the middle of the parietals and thence descends gradually, in an almost straight line to the anterior nares, except for a very slight concavity, or "dishing," of the nasals about the middle of their length.   *Vulpes* has a similar upper profile of the skull, but, because of the higher occiput and more capacious brain-case, the posterior rise is not so steep and the concavity of the nasals is more distinct.   In very strong contrast to the cranium of *Daphoenus*, the sagittal crest is low and weak and is confined to the parietals; at the coronal suture the crest divides into the temporal ridges, which enclose a narrow sagittal area, flat as in *Vulpes*.   The cranium, though slender, elongate and narrowing anteriorly, is relatively more capacious than in *Daphoenus* and the post-orbital constriction is much nearer to the orbits, having the position seen in the existing foxes.   The muzzle, or rostrum, is short and slender and tapers gently forward, not being so abruptly constricted at the line of the infraorbital foramina as it is in *Daphoenus*.

The occiput is low, broad at the base and narrowing toward the summit more than in *Vulpes*, or *Urocyon*, but less than in *Canis;* the crest of the inion is low and weak, very much less prominent than in *Daphoenus*.   In the middle line of the occipital surface there is a low, convex ridge, caused by the vermis of the cerebellum.   The low and broad exoccipitals end infero-laterally in short paroccipital processes, which are pointed and project so directly backward, that they are widely removed from the auditory bullae.   The supra-occipital is relatively large, but does not form so much of the cranial roof as in such modern genera as *Canis*, *Vulpes* and *Urocyon*.   A larger area of the mastoid is exposed on the surface of the cranium than in most Recent genera and the mastoid process is low and inconspicuous.

The *tympanic* forms a large, inflated auditory bulla, which is a single undivided chamber, though in exceptionally well-preserved skulls, a very shallow groove passes diagonally across the ventral surface of the bulla and suggests the presence of an ento-tympanic.   The auditory meatus is a large, irregular hole, without projecting lip and with the principal diameter directed dorso-ventrally.   Proportionally, the bulla is as long antero-posteriorly as in Recent genera (*Canis*, *Vulpes*, *Cerdocyon*, etc.) but is narrower and, therefore, less capacious and it encroaches less on the basi-occipital.

The *parietals* are relatively very large and form most of the cranial walls and roof; the low and thin sagittal crest is a slight prominence at the concavity of the brain-case roof between the occipital crest and the hinder side of the brain-case, a concavity which is shorter and much shallower than in *Daphoenus*, but longer and better defined than in any of the Recent genera.

The *frontals* enclose even less of the cranial cavity than in *Canis* and correspondingly less than in *Urocyon*, or *Cerdocyon;* the forehead is slightly convex. The postorbital constriction is deep and abrupt, which makes the postorbital processes conspicuous, though, properly speaking, they are very short.   Anteriorly, the frontals are deeply incised, to receive the slender nasals, and have long nasal processes, which, however are widely separated from the premaxillae.   There are no frontal sinuses, a difference from *Daphoenus* which, in this respect at least, agrees with Huxley's thoöid division of the Canidae, while *Pseudocynodictis* agrees with the alopecoids, or foxes.

The *squamosal* forms a relatively limited part of the cranial wall; a projecting shelf, narrower than in *Daphoenus*, wider than in *Canis* and other modern genera, extends from the base of the zygomatic process to the post-tympanic process.   The zygomatic process is like that of the wolves, though straighter, extending to the postorbital process of the jugal.

The *jugal* is like that of *Canis*, though with some differences; it is relatively shorter than in the modern genera, not extending so near to the glenoid cavity, and the post-orbital angulation (it can hardly be called a process) is even less conspicuous.   The masseteric surface is broader and more lateral in position and is bounded dorsally by a distinct ridge; the maxillary process is shorter and the frontal, or ascending, process is more slender, but extends farther up the anterior margin of the orbit.   As a whole, the zygomatic arch has nearly the same relative length as in the Coyote, *Canis latrans*, but pursues a more horizontal course, not arching upward so decidedly, in which respect it is more as in *Urocyon*. In view of the more elongate cranial region, one might expect to find correspondingly long zygomatic arches, but the longer cranium is compensated for by the more anterior origin of the zygomatic process.   The *lachrymal* forms but a very small part of the orbital rim and carries a vestigial spine; within the orbit the bone is relatively larger than in *Canis* and the foramen is farther from the frontal suture.   .

The *nasals* are slender, splint-like bones; they resemble those of *Vulpes*, but are proportionately shorter and less dished in the middle.

The *premaxillae* are small and the alveolar portion is weak, in correlation with the minute incisors, and the grooves for the reception of the lower canines are much shallower than in *Daphoenus*.   The ascending ramus is long and slender, but does not reach the frontal; the anterior nares are small, oval in shape and have an oblique position.   The palatine processes are short and very slender and the incisive foramina are small.   This part of the bony palate has an entirely different appearance from that found in *Daphoenus;* the premaxillaries are not extended nearly so far in front of the canines; the incisive foramina are shorter and have no such grooves extending forward from them; the spines are very slender and much shorter, reaching only to the line of the canines and not to that of p$\underline{1}$, as they do in the larger animal.   In most of these respects *Daphoenus* is more like the Recent Canidae than is *Pseudocynodictis*.

The *maxillaries* are relatively very short, especially the preorbital portion; except for the swelling produced by the root of the canine, the facial surface is simply convex, without *fovea maxillaris*.   As in *Daphoenus*, the infraorbital foramen is near the orbit, while in

Canis, but not in *Urocyon*, it is farther removed from the eye, though over p3 in both instances. The palatine processes are short and narrow. The *palatine* bones have nearly the shape and proportions seen in *Canis latrans*, but the palatine notches are more deeply incised, almost as much as in *Urocyon*. The *pterygoids* end in longer, thicker and more conspicuous hamular processes than do those of Recent genera, in some of which, such as in *Urocyon*, the processes are minute, or absent.

The *mandible* has a slender, elongate horizontal ramus, the ventral border of which is regularly and gently convex, becoming concave beneath the ascending ramus; the angular process is a short, blunt hook. The ascending ramus has a stronger backward inclination than in *Canis* and is relatively broader; the masseteric fossa is very deep.

The *cranial foramina* are very minute and hence are frequently obliterated by pressure; they are characteristically cynoid, save in the separation of the carotid canal from the posterior lacerated foramen. The condylar foramen is very small, hardly more than a pinhole, and perforates the ridge which runs mesially from the paroccipital process. Because of the relatively longer auditory bulla, the *foramen lacerum posterius* and the stylo-mastoid foramen are smaller and more concealed than in *Canis*. A real difference from Recent dogs consists in the presence of a separate external opening of the carotid canal, which grooves the inner side of the bulla. The other foramina are essentially cynoid in number and position, as is the whole structure of the cranial basis; nothing suggests viverrine relationships.

BRAIN

The brain is relatively smaller and the cerebral convolutions are much simpler than in the existing Canidae. The olfactory lobes are large and are not covered by the hemispheres, with which they are connected by short, thick olfactory tracts. The hemispheres are pear-shaped, broad behind, tapering forward and decreasing in vertical as much as in transverse diameter. The frontal lobe is short, narrow and vertically shallow; the parietal lobe, which is much larger in every dimension, is demarcated from the frontal by a transverse groove. The temporo-sphenoidal lobe is well developed and adds much to the vertical depth of the cerebrum. Posteriorly, the hemispheres slightly overlap the lateral lobes of the cerebellum, but leave the vermis uncovered.

The sulci which demarcate the cerebral convolutions are few, simple and short, though, almost certainly, the brain-cast does not reproduce all of the sulci. The crucial fissure is not shown in the brain-cast of any Tertiary carnivore that has, as yet, been figured or described, yet it can hardly be doubted that this sulcus, so universal in the brains of Fissipedia, must have been present in the Eocene ancestors of the various carnivorous families. In existing members of the Canidae, the cerebral convolutions are numerous and complex and the sulci pursue a curved course, giving to the convolutions, when the brain is viewed from one side, the appearance of a succession of U-shaped, concentric coils curving around the Sylvian fissure and other sulci are visible, when seen from above.

In all of the Oligocene and lower Miocene Canidae, the brain-casts of which are known, *Daphoenus*, *Pseudocynodictis* and *Cynodesmus*, the sulci are few, shallow and short. In *Pseudocynodictis* but two fissures are visible in the dorsal aspect of the hemisphere, the lateral and the supra-sylvian. The lateral sulcus shows the beginning of a posterior downward curvature, which is very faint on the temporo-sphenoidal lobe. The supra-sylvian fissure is still more curved and this curvature represents a distinct advance over the brain of *Daphoenus.*   The Sylvian fissure is but faintly indicated on the cast, but the rhinal sulcus is very distinct and extends for nearly the whole length of the cerebrum.   The hemispheres of *Pseudocynodictis* retain characters which are now embryonic and transitory.

The *cerebellum* which is separated from the cerebral hemispheres by a bony tentorium, is relatively large and is somewhat more exposed than in the Recent dogs.   The vermis is narrow and prominent and is plainly divisible into three lobes, which presumably correspond to those of *Canis;* the vermis is less regularly curved than in the latter genus, the dorsal and posterior surfaces meeting at nearly a right angle.   The lateral lobes of the cerebellum have a different appearance from those seen in existing members of the family; the *lobus quadrangularis* is narrower and the *lobus lunatus inferior* is very imperfectly developed, while the *lobi semilunares* are either very small, or entirely absent.   A small additional lobe, not represented in *Canis,* lies upon the dorsal surface of the *lobus quadratus* and near the vermis.   Complicated as it appears to be, the cerebellum of *Pseudocynodictis* is really much simpler than that of *Canis,* but is, nevertheless, more complex than that of *Daphoenus.*

THE AXIAL SKELETON

The number of vertebrae has not been ascertained with certainty, but was presumably the same as in *Daphoenus:* C 7, D 13, L 7, S 3, Cd ? 25; the John Day species of the genus are represented by more complete material than those of the White River.

The *atlas* is rather more canine in character than is that of *Daphoenus,* having a short, broad body and moderately developed transverse processes.   The anterior cotyles are shallower and more depressed than in *Canis;* the neural arch is broad antero-posteriorly and quite smooth, without ridges, or tubercles; the inferior arch is very slender and has but a vestigial hypapophysial tubercle.   The transverse processes are rather small and are much less extended antero-posteriorly than in *Canis,* not reaching so far behind the posterior cotyles, nor so far forward upon the neural arch.   In consequence of this, the atlanteo-diapophysial notch is less deeply incised.   The posterior opening of the canal for the vertebral artery presents backward, as it does in *Daphoenus,* but has shifted somewhat to the dorsal side.

The *axis* is more like that of *Viverra* than of *Canis.*   The centrum is long, narrow and much depressed anteriorly, becoming somewhat thicker vertically toward the hinder end, which has a transversely oval and nearly flat face.   The cotyles for the atlas are low and wide and project much less outside of the pedicles of the neural arch than they do in *Canis* and are more convex than in that genus.   The odontoid process is long and slender, more so than in *Viverra,* and the articular surface on its ventral side is not, as in *Canis,* continuous with the anterior cotyles, but is separated from them by a low ridge.   The transverse processes, which are very thin and compressed, are of no great length; the vertebrarterial canal, which perforates the transverse process, is relatively longer than in *Canis.*   The neural spine resembles that of the latter genus much more nearly than it does that of *Daphoenus;* it is long antero-posteriorly, not very high and in front extends far in advance of the neural arch, but behind it does not project beyond the zygapophyses, as it so conspicuously does in *Daphoenus* and the cats, but, as in *Canis,* the dorsal border of the spine curves down into the hinder margin of the neural arch.

The *third cervical vertebra* is markedly different from the same vertebra in *Daphoenus* and much more as in the modern dogs.   The centrum is moderately elongate, depressed and slightly opisthocoelous and has a stout, prominent ventral keel, which is better developed than in *Canis* and ends behind in a tubercle. The anterior face is broad, depressed, convex and very oblique with reference to the long axis of the centrum, while the posterior face is more nearly circular. The transverse process is, in general character, like that of *Canis*, but is relatively less extended from before backward, and less obviously divided into anterior and posterior portions, the ventral border of the process being nearly straight. The relative length of the vertebrarterial canal is much greater than in the modern genus, while the neural arch is long and broad and but slightly convex on the dorsal surface. One noteworthy difference from *Canis* is that the arch does not project over the sides, or pedicles, as an overhanging shelf, or does so but slightly; the neural spine is represented only by an inconspicuous ridge. The zygapophyses are small and project but little in front of and behind the neural arch, a very marked difference from *Daphoenus*. In *Pseudocynodictis*, as in *Canis*, the interspaces between the successive neural arches are very narrow, sometimes hardly visible.

The *fourth cervical* is somewhat shorter than the third, but otherwise, very much like it and like the same vertebra in *Canis*. The transverse process is somewhat larger and heavier than that on the third and the greater breadth of its outer portion makes the arterial canal relatively longer than in *Canis;* the inferior lamella is very thin and light. The neural spine is short and slender, but is relatively better developed than in most modern dogs.

On the *fifth cervical* the neural spine is higher and more slender than that on the fourth.

The *seventh vertebra* may almost be described as a miniature copy of the same vertebra in *Canis*. The transverse process is somewhat longer and thinner than in most species of the existing genus and the neural spine is higher, more slender and pointed.

The number of *dorsal vertebrae* is very probably, but not quite certainly thirteen. The anterior vertebrae of this region have very small, contracted centra, but long and prominent transverse processes and neural spines which are proportionally higher and more slender than in *Canis* and are also inclined backward more decidedly than in the latter. Posteriorly, the centra become longer, broader and more depressed and are quite distinctly keeled in the median ventral line. In addition to this median keel are two shorter and less prominent lateral ridges, which end behind in distinct tubercles, giving a very characteristic appearance to these vertebrae. Posteriorly, the transverse processes become more and more shortened and the neural spines lower and less strongly inclined, but more compressed and antero-posteriorly broadened at the base. The ante-penultimate dorsal, supposedly the eleventh, is the anticlinal vertebra, of which the neural spine is low, broad, compressed and erect. The penultimate (? 12th) and last (? 13th) dorsals are very much like lumbars, with heavy and prominent metapophyses and anapophyses, but no transverse processes, which are present, though small, on the corresponding vertebrae of *Canis*.

The *lumbar vertebrae* almost certainly numbered seven. The lumbar region is proportionately long and stout and, for so small an animal, the individual vertebrae are massively constructed, indicating that there was a powerful musculature in the loins. The centra increase in length up to that of the penultimate vertebra, the first and last being the shortest of the series. These centra are broad and depressed and have distinct ventral medial keels, while the lateral ridges and tubercles are present on the first two lumbars, but not on the last three. The faces are kidney-shaped, slightly convex in front and concave behind, and are placed obliquely to the long axis of the centrum. This obliquity is to provide for the curvature of the loins, which rise to the pelvis, the rump standing considerably higher than the shoulders.   The transverse processes, which are short on the anterior lumbars, increase regularly in length back to the sixth vertebra, on which they are very long; they are slender, depressed, pointed and curved forward.   The neural spines are low, compressed and thin, broad at the base, narrow and pointed at the tip, and are inclined forward rather more strongly than in *Canis*.   Anapophyses are prominent on the anterior lumbars, but diminish posteriorly, becoming vestigial on the fifth, while the metapophyses are conspicuous on all.   The zygapophyses are but moderately concave and convex respectively.

The general aspect of the lumbar region is not canine in character, but rather resembles that of the civets and mustelines.

The *sacrum* is short and consists of three vertebrae, only the first of which supports the ilia.   The first sacral has a broad and much depressed centrum and large, expanded pleurapophyses, which give considerable width to the vertebra.   The neural spine is a low ridge, while the spines of the second and third are separate and higher.   The transverse processes of all the sacrals are fused into continuous lateral ridges, but those of the third vertebra extend outward much farther than the others and end in points, a structure which gives to this sacrum an appearance very different from that of *Canis*.   The pre-zygapophyses of the first vertebra are large and prominent, but all the other zygapophyses of the sacrum are small, and the foramina for the passage of the spinal nerves are remarkably small.   The centrum of the last vertebra is almost as large as that of the first and the widely extended transverse processes make the sacrum of nearly uniform width.

No complete tail has yet been obtained, but many *caudal vertebrae* have been found, sufficient for a reconstruction, and these show that the tail was very long and relatively thick, longer and stouter than in any of the Recent Canidae and more as in such long-tailed viverrines as *Herpestes*.   The anterior caudal vertebrae have short, heavy centra and very long, broad and depressed transverse processes, which extend out nearly at right angles with the axis of the centrum, making the first caudal about equal in breadth to the last sacral.   The zygapophyses are large and prominent on the anterior caudals, and these are followed by a number of vertebrae with very elongate centra, which resemble, in miniature, the corresponding ones of *Daphoenus* and have distinct vestiges of the various processes.   Toward the tip of the tail the vertebrae become very slender and of a cylindrical shape, the centra being slightly contracted in the middle and expanded at the ends.

The *ribs* are remarkable chiefly for their length and slenderness and their sub-cylindrical shape.   Tubercles appear to be absent from the twelfth and thirteenth pairs.   The thorax would seem to have been narrow and compressed, but relatively deeper and more capacious than in *Daphoenus*.   The *sternum* is of the usual Carnivorous type, without being especially like that either of the dogs or of the civets.   The manubrium is longer, narrower and more compressed than in *Canis*.   The first pair of ribs are attached to two wing-like processes, which are unusually far from the second pair.   In front of the processes the bone is compressed and very narrow and, for much of its length, the manubrium has a ventral keel. The segments of the mesosternum are more elongate, more slender and depressed and more contracted in the middle than are those of the Recent Canidae.

The *baculum* is very perfectly preserved in a specimen figured and briefly described by Hatcher, who says of it: "This bone is grooved throughout nearly its entire length.

The distal extremity is not bifurcated as in *Daphoenus*, but is solid, abruptly curved and terminates in much the same manner as in *Procyon lotor*" (p. 83). Flower ('69) has pointed out the characteristics of the *os penis* in the three sections into which he divides the Fissipedia. The Arctoidea "all have a large penis with a very considerable bone, which is usually more or less curved, somewhat compressed, not grooved, dilated posteriorly and often bifurcated, or rather bilobed in front" (p. 14). The cats and viverrines "all have a comparatively small penis, with a more or less conical termination and of which the bone is small, irregular in shape, or not unfrequently altogether wanting" (p. 22). To this statement *Cryptoprocta* is an exception, having a bone which is relatively long, "slender, compressed, slightly curved, not grooved or divided anteriorly, rounded and slightly dilated at each end, but thickest posteriorly" (p. 23). In the hyenas the bone is wanting. The dogs resemble the raccoons, mustelines etc. in having a large baculum, "though the os is of a different form, being straight, wide, depressed and grooved" (p. 26).

Both *Daphoenus* and *Pseudocynodictis* depart widely from existing dogs in the development of the baculum and agree much better with the raccoons and mustelines, though not entirely like any existing form of the latter. In *Pseudocynodictis* the bone is long, slender, compressed and grooved on the dorsal and lateral aspects; the sigmoid curvature is decidedly more pronounced than in *Daphoenus* and the distal end is not bifurcate, as in the latter, nor perforated, but is slightly broadened and knobbed. The un-canine character of the baculum in the Oligocene dogs seems to imply that they retained the form which was originally common to all the fissiped families and which still persists in the mustelines and raccoons and survives, with no radical change, in the modern dogs. In the cats and their allies, the civets and hyenas, the bone has been entirely lost, or reduced to a vestigial condition, though to this statement *Cryptoprocta* is an exception. It seems very much more likely that this bone should have been retained in that phylum of the civet family which terminated in *Cryptoprocta*, than that the latter should have acquired it independently.

From the seemingly arbitrary and capricious occurrence of the baculum among the fissiped genera, it may be inferred with considerable confidence, that the Eocene ancestors of all the existing families, presumably the Uintacyonidae, possessed the bone of large size and curved form and that in subsequent time it underwent various modifications in the various families, the "Arctoidea" retaining it in its almost original form, the "Aeluroidea" getting rid of it more or less completely. If this interpretation is correct, it will serve as a key to unlock some of the mysteries of the evolutionary process. It is true that this bone has not been found in connection with any Eocene skeleton, but it is very rarely found at all. Only three examples of it have been reported from Tertiary beds.

THE FORE LIMB

The *scapula* is remarkable and in character it is viverrine or raccoon-like rather than canine. The blade is rather short proximo-distally and is divided by the spine into pre- and post-scapular fossae of nearly equal breadth, while in the modern dogs the scapula is long, narrow and of sub-quadrate shape; the spine pursues an oblique course and is so placed as to make the post-scapular fossa much the larger of the two. The glenoid cavity is moderately elongate antero-posteriorly and narrow transversely. The coracoid process is unusually large and forms an incurved hook, which, however, does not appear prominently when the shoulder-blade is viewed from the external side; in Recent dogs the coracoid is reduced to much smaller proportions. A resemblance to *Canis* is seen in the broad neck of the scapula and in the absence of any well-defined coraco-scapular notch. The coracoid border is slightly concave at the neck and then curves forward and upward; the glenoid border is, as usual, straight and steeply inclined. The spine is high and ends in a very long and prominent acromion, which descends below the level of the glenoid cavity. Such an acromion suggests that in this genus the clavicles were much more complete than in existing dogs. A very large metacromial process. is also present. The metacromion is found in most of the existing families of Fissipedia, but it is seldom so large as in the present genus; among Recent genera, the nearest approach is perhaps *Arctictis*.

The *humerus* is much more suggestive of viverrine than of canine affinities; it is longer than the forearm, but short in proportion to the length of the back. The head is strongly convex and projects more behind the plane of the shaft than in modern dogs; the external tuberosity is a heavy, but low ridge, which barely conceals the head, when the humerus is seen from the front; a large, irregularly circular area, near the hinder end of this ridge plainly indicates the place of insertion of the *infra-spinatus* muscle. The external tuberosity is both lower and shorter than in *Canis*, but the internal one is rather more prominent and the bicipital groove is more widely open, more internal in position and more of it is visible from the anterior side.

When seen from the side, the shaft exhibits a sigmoid curvature, which is somewhat better marked than in *Canis*, and, for most of its length, it is laterally compressed and has but a very short cylindrical portion before widening at the distal end. Most of the ridges and prominences for muscular attachment are better developed than one would expect in so small an animal. The deltoid ridge is much more prominent than in existing dogs and more like that of the cats and viverrines; the supinator ridge is likewise very much more conspicuous than in *Canis*, in correlation with the power of rotation of the radius which *Pseudocynodictis* appears to have retained in almost undiminished degree. On the other hand, the rough ridge (*spina humeri*) which runs down the outer side of the shaft from the head and serves for the attachment of the *teres minor, anconaeus externus* and *brachialis internus* muscles, is much fainter than in *Canis* and the *linea tuberculi minoris* is very feebly marked. The supra-trochlear fossa is shallow and the anconeal is much smaller and shallower than in *Canis* and there is no perforation of the shaft. The internal epicondyle is much more prominent and relatively more massive than in the Recent genus and the foramen is a long, narrow slit; the external epicondyle is smaller than in the latter.

The humeral trochlea has a much smaller proximo-distal diameter than in Recent Canidae, preserving a more primitive character and resembling the trochlea of such viverrine genera as *Viverra* and *Cynogale*. The radial surface is small and simply convex, while the ulnar facet is much larger than in *Canis* and its inner flange is more produced distally and forms a sharper edge.

The *radius* is not at all suggestive of canine relationships and rather resembles that of the cats and civets. The capitellum is small and of sub-discoidal shape; while it is somewhat more extended transversely than in *Felis*, it is much less so than in *Canis;* the articular surface is moderately concave and is slightly notched on the anterior border. The proximal facet for the ulna is a simple, convex band, separated from the humeral surface by a distinct angle and altogether like that of *Daphoenus*. The mode of articulation at the elbow-joint and the large development of the supinator ridge imply that in the present genus a considerable degree of freedom of rotation had been preserved, though probably less than in the cats and in many viverrines. The bicipital tubercle is prominent, but occupies a more posterior position than in either cats, or viverrines and is not visible from the front.

The shaft of the radius is relatively short, slender and rounded, very different from the broad, oval and antero-posteriorly compressed shaft seen in *Canis;* it has a slight double curvature, arching forward and outward, and is of almost uniform breadth throughout, except at the distal end, where it broadens considerably. A striking difference from *Canis* consists in the great size and prominence of the styloid process, which forms a relatively enormous tuberosity (if one may use such a word in connection with so small a creature); it is even much larger proportionally than in the cats and civets and is as large as in *Mellivora*, though of a different shape. In *Daphoenus* the styloid process is very prominent and of feline character, but is not so large as in the present genus. The radius figured by Schlosser and by him attributed to the European *Cynodictis* ('89, Taf. VII, Fig. 8), has a styloid process in the form of an immense, recurved hook, much longer and more slender than in the American genus and of an entirely different appearance. The distal tendinal sulci are not very well-marked, though that for the abductor and extensor muscles of the pollex is a deep groove. The distal facet for the ulna is smaller and less deeply impressed than in *Canis;* that for the carpus is small and slightly concave, narrowing toward the internal side; it does not extend over upon the styloid process, from which it is separated by a broad and deep notch.

The *ulna* is, in its way, as peculiar as the radius. The olecranon is typically fissiped in form and differs from that of the creodonts in its comparative shortness and breadth; though relatively somewhat longer than in *Canis*, it is hardly so long as in *Daphoenus*. The sigmoid notch is not quite so deep as in the former and, in particular, the internal facet for the humerus projects less in front of the shaft and the external process is very weak. The radial facet is narrower and less concave than in modern Canidae, but has a somewhat greater vertical diameter. The shaft is decidedly less reduced than in Recent members of the family and, for most of its length, is little, or not at all, more slender than that of the radius. In its proximal portion, the shaft is much more compressed laterally and thicker antero-posteriorly than in *Canis*, in which this part of the shaft is trihedral. The middle and distal portions in *Pseudocynodictis* are trihedral, none of it having the sub-cylindrical shape which characterizes the Recent genus. The distal end has quite a different shape from that of *Daphoenus*, a difference which is due to the much greater prominence of the radial facet in the latter. The carpal facet is very small.

The *manus* is small and weak. The *carpus* is thoroughly fissiped, but peculiar in certain details. A *scapho-lunar*, formed by the coalescence of the scaphoid, lunar and central, is present and, in general, resembles that of *Canis;* it is very short proximo-distally, especially on the radial side, where it thins almost to an edge. The radial surface is simply convex both transversely and palmo-dorsally and has not the saddle-shaped extension at the interno-palmar angle which appears in *Canis*. This proximal articular surface descends low upon the dorsal side of the bone, as in the modern plantigrade and semi-plantigrade carnivores. The hook-like process which arises from the postero-internal angle of the scapho-lunar is much smaller and less massive than that of *Canis*. Another difference from the latter is the absence of any distinct facet for the pyramidal, the surface for the radius and that for the unciform almost meeting along the ulnar side of the bone.

On the distal side of the scapho-lunar are four facets, for all of the carpal elements of the distal row: that for the unciform is relatively smaller than in *Canis* and is confined to a narrow band near the ulnar border; the surface for the magnum is much the same as in the latter, but is somewhat more oblique; the facet for the trapezoid is fairly large and keeps more nearly parallel with that for the magnum than in the Recent dogs, while the surface for the trapezium is small and of almost circular shape.

The *pyramidal* is a very different-looking bone from that of the Recent Canidae, being broad, depressed and scale-like in shape; its proximo-distal diameter is very small and proportionately much less than in *Canis* and there is no such process from the ulnar side of the bone as in the latter, in which the pyramidal articulates with the head of the fifth metacarpal by a much more extensive surface than in *Pseudocynodictis*, in which the pyramidal is very much like that of the Recent Viverridae. The proximal side of this bone is divided into two narrow and somewhat concave facets for the ulna and pisiform respectively, of which the latter is slightly the longer. On the distal side is a single large surface for the unciform and, on the palmar side of this, a very narrow surface which appears to articulate with the head of the fifth metacarpal.

The *pisiform* differs very decidedly from that of *Canis*. This bone is small and light; its proximal end is much depressed and very broad (in the modern genus the principal diameter of the proximal end is the vertical one) and the surfaces for the ulna and the pyramidal are correspondingly broadened transversely and narrowed vertically. The pyramidal facet is the larger of the two and is rather deeply concave, while that for the ulna is nearly plane; together, the two surfaces form an acute angle and are separated only by an inconspicuous ridge. The distal end of the pisiform is moderately expanded, but in the vertical dimension, so that the proximal and distal expansions are almost at right angles with each other. Between the two expansions, the bone is much contracted and very slender, which is in strong contrast to the shape in *Canis*.

The *trapezium* is very small and differently shaped from that of *Canis;* its principal dimension is the dorso-palmar one, and the transverse diameter is the least. The surface for the scaphoid, which in the existing genus is a very oblique, convex facet, is in *Pseudocynodictis* entirely proximal in position and nearly plane, and, on the ulnar side there is no such large, concave facet for the trapezoid as occurs in Recent dogs; the distal surface for the head of the first metacarpal is less distinctively saddle-shaped than in the latter. In view of the stout and well-developed pollex, the small size of the trapezium is somewhat surprising.

The *trapezoid* is shaped very much as in *Canis*, but with certain minor differences, especially in the very small proximo-distal diameter and in the thinning of the bone to an edge on the ulnar side. The proximal end bears a simply convex surface for the scapholunar; while the distal surface, for the second metacarpal, is very slightly saddle-shaped; on the palmar side, the trapezoid contracts to a blunt point.

The *magnum* is small and the dorsal face, when all the carpals are in their natural positions, is minute, especially the proximo-distal dimension. In shape, the magnum does not differ much from that of *Canis*, but the proximal surface is narrower and rises more abruptly to the head and, on the palmar side, the bone broadens out in a manner not repeated in *Canis*. The unciform facet is large and plane and does not rise so high upon the head as in the last-named genus. On the radial side there is no distinct facet for the trapezoid, but there is a well-defined surface for the projection from the head of the second metacarpal, which is proportionately larger than in the modern dogs. On the distal end of the magnum is a narrow surface for the third metacarpal, a surface which is less concave palmo-dorsally than in the latter.

The *unciform* is viverrine rather than canine in character, being much narrower in proportion to its proximo-distal length than in Recent Canidae. The facet for the scapho-lunar, which in *Canis* has an almost entirely proximal position, is in the present genus much more nearly lateral. The pyramidal facet also is decidedly more steeply inclined, the two articular surfaces meeting at an acute angle and making the proximal end of the unciform narrow and wedge-shaped. On the radial side there is a large surface for the magnum and a smaller one, confluent with it, for the extension from the head of the third metacarpal. The distal facets for the fourth and fifth metacarpals are relatively narrower than in *Canis*.

The *metacarpals*, five in number, are remarkably short, slender and weak, with but little resemblance to those of *Canis*.

The *first metacarpal*, though very small, is yet not nearly so much reduced as in the modern genus, taking the length of mc. III in each as the standard of comparison. The head is relatively thicker and heavier than in *Canis* and, on the radial side, internal to the trapezium facet, is a tubercle for the attachment of the lateral ligament; the facet itself is much less deeply concave transversely than in *Canis*, but more convex palmo-dorsally. The shaft is short, slender, arched toward the dorsal side and of oval cross-section, tapering considerably toward the distal end. The distal trochlea is very small, of nearly hemispher-ical shape and has a distinct palmar keel.

The *second metacarpal* is incompletely known, but is seen to be much stouter than the first and about equal to the fourth in diameter. The head is narrow and bears a saddle-shaped surface for the trapezoid, but sends out a projection from the ulnar side, which over-laps mc. III and abuts against the magnum.

The *third metacarpal*, though short and slender, is yet the longest and stoutest of the series and has a viverrine, rather than a canine appearance; the ulnar projection from the head, slightly overlaps that of mc. IV, and articulates with the unciform. The shaft differs in shape from that of *Canis* in retaining the oval cross-section and is of almost uniform diameter throughout, though broadening somewhat at the distal end. The distal trochlea is more canine than in *Daphoenus*, being broad and semi-cylindrical in shape rather than sub-spherical.

The *fourth metacarpal* is imperfectly preserved, but appears to have had about the same length as mc. III and to have formed with it a symmetrical pair, though these are not closely appressed, but diverge somewhat distally. The head has a simply convex facet for the unciform and is relatively somewhat narrower than in the modern dogs owing to the overlapping of the head by mc. III, in order to reach the unciform.

The *fifth metacarpal* is remarkably short, much shorter proportionately than in *Canis;* the head is less broadened and thickened than in the latter and bears a simple, convex facet for the unciform. In the modern genus this metacarpal has, in addition, a large facet for the pyramidal, which extends down over the unciform to a contact with mc. V and this seems to be represented in an obscure and limited way, in *Pseudocynodictis*. The shaft is slender proximally and broadens distally, the reverse of the proportions in *Canis*, and the distal trochlea is small and of a more hemispherical, less semi-cylindrical shape than in the latter.

The *phalanges* are still incompletely known.   The proximal phalanx of a median digit is short, slender and straight and is relatively broader and more depressed than in *Canis;* the distal end is shaped as in *Daphoenus*.   The *unguals* differ in certain not unimportant respects from those of both of the genera named and are, on the whole, intermediate between them.   As compared with the unguals of *Daphoenus*, they have less concave proximal articulations, a smaller subungual process and a much less extensive bony hood, which, in life, was reflected over the base of the horny claw; indeed, there is hardly any of the hood. Compared with those of *Canis*, these unguals are decidedly thinner, more acutely pointed and have more concave proximal ends.

The Hind Limb

The *pelvis* approximates more nearly the modern canine type than does that of *Daphoenus*, though still retaining a number of primitive characters.   A conspicuous difference from the Recent members of the family is the elongation of the post-acetabular part of the pelvis and in the consequent change of shape of the obturator foramina.   The ilium is moderately elongate and, in shape, rather more viverrine than canine, the peduncle is short and laterally compressed, but of considerable dorso-ventral breadth.   The anterior expansion is relatively less than in *Canis*, in which the ilium widens gradually to the crista, while, in *Pseudocynodictis* it attains nearly its full breadth immediately in front of the peduncle and, from that point forward, the ischial and acetabular borders are almost parallel; the shape is very much as in *Herpestes*.   The gluteal surface is not the broad and simple concavity of *Canis*, but, as in *Daphoenus* and *Dinictis*, there is a narrow dorsal depression and beneath this a convex ridge, but the ridge is not so prominent as in the White River genera mentioned.   The iliac surface is short and narrow and the sacral surface is small and placed far back.   Seen from above, the two ilia are less everted and divergent than in *Canis*.   The acetabular border ends in a tubercle and the ilio-pectineal process is also prominent.

The ischium is relatively long and its anterior portion is slender and, posteriorly, it expands into a broad plate, which is more vertical in position, less everted and depressed than in *Canis* and the ischial tuberosity is much less prominent; on the other hand, the spine of the ischium and the sciatic notch are more distinct and are placed farther behind the acetabulum, though relatively not so far as in *Herpestes*.   The obturator foramen is narrower and more elongate than in Recent dogs and its anterior border is notched by the obturator sulcus.   The anterior, or descending ramus of the pubis is long and slender and, with its fellow of the opposite side, encloses a broad anterior pelvic opening.   The horizontal ramus is proportionally longer and stouter and the symphysis is longer than in the modern dogs, almost as long as in the cats; this ramus is less flattened and depressed than in the former and is a prominent ridge along the ventral side of the symphysis.

The *femur* is long and slender and, in essentials, differs little from that of *Canis*.   The head is small, of hemispherical shape and is set upon a somewhat longer and more distinct neck than in the Recent genus and projects more directly inward, less proximally; the pit for the round ligament is deeply impressed, but very small.   The great trochanter is lower than in *Canis* and is separated from the head by a narrower, shallower notch and the digital fossa is smaller.   The second trochanter is much as in the latter, but rather more slender and pointed and the inter-trochanteric ridge is better developed, especially in the larger individuals. What may be regarded as a vestige of the third trochanter is a low, short, thick and rugose ridge, which, as in *Daphoenus*, is placed a little below the great trochanter. The shaft of the femur is long, slender, arched strongly forward and, as might be expected in so small an animal, the ridges for muscular attachment are less developed than in the Recent species. On the anterior face no ridge for the *vastus externus* muscle is discernible and on the hinder side the *linea aspera* is neither so long nor so prominent as in *Canis*. The distal end differs from that of the latter in the smaller size and less prominent projection of the condyles and trochlea. The trochlea resembles that of the civets in being shallow and having the borders of nearly equal prominence and in the absence of a distinct supra-patellar fossa. Fabellae were evidently, as in the existing species of the family, attached to the proximal faces of the condyles.

The *patella* is herpestine rather than canine in character. It is a short, rather wide, thin and scale-like bone, of sub-quadrate rather than oval shape. The articular surface for the femur, in correlation with the shallowness of the rotular groove, is but slightly concave proximo-distally and even less convex transversely.

The *tibia* is of about the same length as the femur and is much longer than the radius and differs from that of modern canids in several particulars. The proximal condyles are of nearly equal size, but the external one projects much farther behind the plane of the shaft than in *Canis* and on the distal side of the overhanging shelf so formed is a facet for the head of the fibula, which is larger and more distinct than in the modern genus. The tibial spine is low and bifid, but the two parts are closely approximated, as the inter-condylar notch of the femur is narrow. The cnemial crest, though heavy and prominent, is less so than in *Canis* and the sulcus for the *extensor longus digitorum* is less deeply incised.

The shaft of the tibia is proximally stout and trihedral, but, for most of its length it is slender and sub-cylindrical, broadening moderately at the distal end and it has a double curvature, arching forward and outward. The various ridges for muscular attachment are much the same as in *Canis* and more conspicuous than those on the femur. The distal articular surfaces of the tibia are intermediate in character between those of *Daphoenus* and those of *Canis*. The grooves for the astragalar condyles are deeper and the inter-condylar ridge higher than in the former, less so than in the latter, and the articular surface is not invaded by a transverse sulcus. The internal malleolus is relatively smaller than in *Daphoenus*, but, as in that genus, it forms a heavy, prominent ridge, which extends across the whole dorso-plantar diameter of the bone, while in *Canis* the process has not half of this extension. The groove for the tendon of the long flexor muscle is very distinct and has more elevated borders than in modern dogs. The distal fibular facet is somewhat larger than in the latter and differs in having its principal diameter transverse. The resemblance in the form of the distal end of the tibia between *Pseudocynodictis* and *Daphoenus*, on the one hand, and the primitive sabre-tooth *Dinictis*, on the other, is very marked and very suggestive. Among living forms, the tibia of the viverrine genus, *Herpestes*, offers a close analogy to that of the genera named.

The *fibula* is less reduced than in existing Canidae and both the shaft and the ends are larger. The proximal end is relatively larger and heavier than in *Canis* and, though smaller than in *Dinictis*, it has a very similar shape; its principal diameter is the antero-posterior one; transversely it is narrow and the thickening of the anterior and posterior borders is much less than in the latter. The facet for the head of the tibia is large, subcircular in shape and proximo-lateral in position. The shaft, though slender and delicate, is relatively far less so than in *Canis*. Tibia and fibula are in contact only at the two extremities. The distal end is expanded and thickened to form a stout external malleolus, which is somewhat smaller than in *Dinictis*, yet has a very similar shape and has on its outer side a sulcus for the peroneal tendon. The distal tibial facet is a narrow band. There is no articulation with the calcaneum.

The *pes* in general appearance reminds one of the Viverridae. The *astragalus* is very like that of *Daphoenus* on a small scale, but with some resemblances to *Canis;* the proximal trochlea is but little more deeply grooved than in the former and is, therefore, shallower than in the latter, but its borders have the same clean-cut angularity as in the modern species, instead of curving gradually into the lateral surfaces for the tibial and fibular malleoli. In *Canis* the trochlear articular surface is continued over upon the neck of the astragalus, but not in either genus of White River canids of which the pes is known. The neck of the astragalus is relatively longer than in the modern dogs, or even than in *Daphoenus*, being more as in such viverrine genera as *Paradoxurus*. The head, with its convex navicular facet, is shaped as in *Canis*, except that it is more depressed planto-dorsally. In *Daphoenus* there is a distinct facet for the cuboid, which meets the navicular facet almost at right angles; in *Pseudocynodictis* this cuboidal facet is much smaller and, in some individuals is altogether wanting, while in *Canis* astragalus and cuboid are not in contact.

As in *Daphoenus*, the external surface for the calcaneum is more oblique in position and more simply convex than in *Canis*, but the sustentacular surface is different from that of both of the genera named; it agrees with that of *Daphoenus* in being shorter and wider than in modern species of the family, but, as in the latter, this facet is confluent with that for the navicular. The inter-articular sulcus is somewhat deeper than in *Daphoenus*, but shallower than in *Canis*. In the latter there is a third calcaneal facet on the astragalus, which forms a narrow band upon the fibulo-plantar side of the head and is connected at one end with the surface for the sustentaculum. This accessory calcaneo-astragalar articulation is not found in either of the White River genera.

The *calcaneum*, like the astragalus is more viverrine than canine in general appearance and quite closely resembles that of *Paradoxurus*, but the likeness to that of *Daphoenus* is even closer. The tuber is slender and compressed and is proportionately shorter than in *Canis*, in which the tuber makes up more than two-thirds of the whole length of the bone, while in *Pseudocynodictis* it forms but two-fifths of the length. The free end of the tuber is moderately thickened and club-shaped and is deeply grooved by the tendinal sulcus. As in *Daphoenus*, the dorsal and plantar borders of the tuber are almost parallel, making the diameter nearly uniform, not increasing distally, as it does in *Canis*. On the fibular side of the calcaneum, near the distal end, is a very prominent process for the attachment of the lateral ligaments. This prominence does not occur in the Recent Canidae, but is very conspicuous in such primitive carnivores as *Daphoenus* and *Dinictis* and it recurs among such Recent plantigrade and semi-plantigrade fissipeds as *Procyon, Gulo, Paradoxurus;* usually, the process is larger in the modern genera named.

The calcaneal facets for the astragalus differ somewhat from those of both *Daphoenus* and *Canis;* in the latter the external astragalar facet is in two parts, one of which presents distally, the other dorsally, and the two meet at an angle which does not much exceed 90°; in the former the surface is a single, continuously curved convexity, not divided by an angulation. In *Pseudocynodictis* the two parts are distinguishable, as in *Canis*, but they meet at a more open angle. The sustentaculum is of moderate prominence and, as in *Daphoenus* it has a sub-circular facet for the astragalus and is not so oblique with reference to the long axis of the bone, as it is in modern dogs. On the plantar side, between the sustentaculum and the tuber calcis, is a groove, the *sulcus flexoris hallucis*, which is better marked in *Canis* than in either of the White River genera. This is a curious fact, because in both of the latter the hallux was fully developed and functional and in the modern genus it is reduced to the merest vestige. In the latter genus there is a third facet for the astragalus, a small, plane surface, distal to the sustentaculum, from which it is separated by a narrow sulcus; continuous with this accessory facet, but at a right angle with it, is a small facet for the navicular. Neither of these accessory articular surfaces occurs in *Pseudocynodictis*. The facet for the cuboid, which in Recent dogs is almost plane and of semi-circular shape, is deeply concave and nearly circular in outline.

The *cuboid* is relatively long and narrow, differing from that of *Canis* principally in the smallness of its transverse and dorso-plantar diameters. The facet for the calcaneum is, as in *Daphoenus*, more convex than in existing dogs. The hook-like projection from the plantar side, which is very large and prominent in *Daphoenus* and in *Canis*, is even more massive in the present genus; it is inconspicuous and continuous with the projection from the fibular side, which overhangs the deep tendinal sulcus. The astragalar facet is small and is confined to the dorsal side of the cuboid, being less extensive than in *Daphoenus*. The facet for the navicular is not so prominent as in the latter and is continuous with that for the ecto-cuneiform and the distal end of the cuboid is similar in having a concave surface for the head of mt. IV, while that for V is lateral in position. In *Canis*, on the other hand, the facet for mt. IV is almost plane and that for mt. V is entirely distal.

The *navicular* is almost a copy, in miniature, of that of *Daphoenus* and presents the same differences from that of *Canis*. Seen from the proximal end, it is of more regularly oval shape and is less contracted on the plantar side than in the Recent genus and the position of the navicular in the tarsus is different. In *Canis* this bone has been, as it were, somewhat rotated, making the dorso-plantar diameter the principal one, and on the plantar border it has been brought into contact with the calcaneum, for which it has acquired a special facet. It is interesting to note, as Rütimeyer has shown, that a similar, but more extensive rotation of the tarsal elements has occurred in the horses, which are the pre-eminent "cursorial machines." In the White River genera, on the other hand, the principal diameter of the navicular is transverse and, owing to the elongation of the neck of the astragalus, the bone is displaced so far that it can have no contact with the calcaneum, from which it is excluded by the articulation of the astragalus and cuboid. The surface for the cuboid is small and is confined to the dorsal moiety of the navicular.

The *ento-cuneiform* has nearly the same shape as in *Canis*, elongate proximo-distally, but very narrow transversely; the navicular facet is smaller than in the latter and there is no such distinct surface for the meso-cuneiform. The distal end, for mt. I, is more deeply concave. The *meso-cuneiform* is minute and, as in the Fissipedia generally, its proximo-distal diameter is much less than that of the adjoining ento- and ecto-cuneiforms, making a recess in the distal row of tarsals. The only articular surfaces visible on this tarsal are those for the navicular and mt. III.

The *ecto-cuneiform* is the largest of the three. Compared with that of *Canis*, it is narrower in proportion to its proximo-distal length and has a shorter dorso-plantar diameter, but the process projecting from the plantar side is even more prominent and is more thickened and club-like at the free end.   On the tibial side is a minute facet (not double, as it is in *Canis*) for the side of mt. II.   The surface for the cuboid is smaller than in Recent dogs and is confined to the dorsal border, while at the infero-external angle of the bone is a minute facet for the head of mt. IV, which does not appear in *Canis;* the distal surface for mt. III is less concave and less prolonged to the plantar side than in *Canis.*

The *metatarsus* consists of five well developed members, which are longer and stouter than the metacarpals; the disproportion between fore and hind legs and feet is much greater than in the Recent dogs and more as in certain of the Viverridae, such as *Herpestes* and *Paradoxurus.*   The *first metatarsal* indicates a well-developed hallux, though somewhat more reduced than in *Daphoenus* or such recent viverrines as *Cynogale.*   The head has a narrow, convex facet for the ento-cuneiform and, on its tibial side is a large rugose prominence for the attachment of the lateral ligament.   The shaft is very slender and is slightly concave along the tibial border.

The *second metatarsal* is stouter than the first and more slender than the third; the very narrow proximal end is wedged in between the ecto- and ento-cuneiforms; a stout prominence arises from the plantar side of the head.   The shaft is slender and of oval cross-section.

The *third metatarsal* is the stoutest of the series; the head is broad on the dorsal, narrow on the plantar side, where there is a large projecting process, which is more prominent than in *Canis.*   The facet for the ecto-cuneiform is convex (in the Recent dogs it is slightly concave) and oblique in position, inclining downward toward the tibial side.   Deep sulci invade the head on both sides; on the tibial side the sulcus is narrow, but that on the fibular side is broad.   A deep pit on the fibular side of the head receives a corresponding prominence on mt. IV and an additional facet for the same metatarsal is found on the plantar projection, so that the two median metatarsals are very firmly interlocked.   For most of its length, the shaft is of oval cross-section, very different from the quadrate shape which is characteristic of *Canis*, though an approximation to this shape occurs in the proximal portion of the shaft, where mt. III and IV are closely appressed.   The distal end is broadened and planto-dorsally compressed; the trochlea resembles that of mc. III, but is lower.

The *fourth metatarsal* is a trifle more slender than mt. III.   The head is narrow and the surface for the cuboid is slightly convex in both directions and the plantar extension is neither so broad nor so prominent as in *Canis.*   On the tibial side is a rounded protuberance which fits into the depression on the head of mt. III, as above mentioned, and on the fibular side is an excavation for a prominence on mt. V, and, proximal to this is a narrow, but well-defined facet for the same metatarsal.   The proximal part of the shaft, where it is appressed to mt. III has the quadrate shape of Recent dogs.   The digits of the pes diverge less than do those of the manus.

The *fifth metatarsal* we have not seen.

The *phalanges* of the hind foot are considerably larger than those of the fore foot, but are otherwise like them.

RESTORATION

Matthew's figure of the skeleton (Fig. 6) cannot properly be called a restoration, as nearly all the bones are represented, though much scattered, in the fine specimen belonging to the American Museum, and there is very little doubt as to any part of it. All who have examined the skeletal material of these small canids of the White River and John Day stages of the Oligocene have been struck by the differences in proportions and appearance from all Recent Canidae and their far greater likeness to the Viverridae. Cope, who had before him some unusually complete material of the John Day species, *P. geismarianus*, wrote of it: "Although the skull and pelvis of this species have about the size of those of the fisher, (*Mustela canadensis*) the vertebrae and humerus are more slender and the anterior foot is decidedly smaller. It is probable that the *Galecynus* [*i.e.*, *Pseudocynodictis*] *geismarianus* resembled a large *Herpestes* in general proportions rather than a *Canis*. It stood lower on the legs than a fox and had as slender a body as the most 'vermiform' of the weasels, the elongation being most marked in the region posterior to the thorax. The tail was evidently as long as in the Ichneumons. Its carnivorous propensities were as well developed as in any of the species mentioned, although, like all other *Canidae* of the Lower Miocene period, the carnassial teeth are relatively smaller than in the recent types" (*Tert. Vert.*, p. 929).

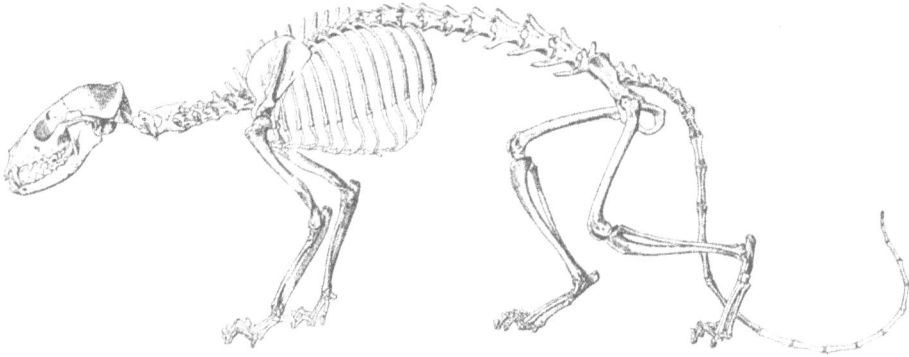

FIG. 6. *Pseudocynodictis gregarius*, skeleton × 5/18. (After Matthew.)

In 1898 the senior author wrote of the White River species, *P. gregarius*, as follows: "The general appearance of the *Cynodictis* [*i.e.*, *Pseudocynodictis*] skeleton has little about it to suggest canine affinities, but has some resemblance to the civets and especially to the herpestine section of that family. This resemblance is not merely a general one of outline and proportions, but may also be traced in many of the details of structure. The small head, with its elongate and narrow cranium and short, tapering muzzle, is of strikingly viverrine character. So is also the neck, which is relatively long and stout, the vertebrae having heavy centra and well-developed processes. The resemblance to the civets continues into the thoracic region, where the vertebrae are small, especially in the anterior portion, and have short, slender neural spines.· . . . The lumbar region is long and is strongly curved upward; the vertebrae are much elongated, with stout depressed centra, very long, slender and anteriorly directed neural spines, which are not like those of modern dogs and civets and most resemble the spines of *Lynx*. The transverse processes are likewise peculiar in their length and slenderness. The tail is unlike that of the modern dogs, being much longer, stouter and in every way better developed; it was not, perhaps, quite so long proportionately as in *Herpestes*, but nearly so. This, however, is a primitive feature, which is common to the greater part of the earlier carnivores and ungulates and is even more conspicuous in *Daphoenus* . . . , while the White River Machairodonts, *Dinictis* and *Hoplophoneus* [*i.e., Drepanodon*] have very long and massive tails" ('98, pp. 398–9).

Dr. Matthew wrote in a very similar strain, though at somewhat less length, and came to similar conclusions.

RELATIONSHIPS OF PSEUDOCYNODICTIS

The upper Eocene and lower Oligocene of Europe, especially the *Phosphorites* of southern France, contain a multitude of small carnivores, canines, mustelines and viverrines, the affinities of which are a labyrinthine welter of confusion. How the small canids of the North American Oligocene are related to those of Europe is exceedingly difficult to determine, and must, for the present, remain an open question. It is clear that in the White River stage we have two series, or tribes of the family Canidae, the larger animals, *Daphoenus, Protemnocyon, Brachyrhynchocyon*, which have the frontal sinus, and the smaller forms typified by *Pseudocynodictis*, which are without the sinus. When the history and ramifications of the family are better known than they are at present, it will assuredly be necessary to recognize several distinct subfamilies and two of these will be the larger and the smaller dogs of the White River stage. Both of these continued into the John Day, after which time the smaller animals disappeared from North America. Wortman and Matthew were of the opinion that certain South American fox-like wolves, which they named *Nothocyon* (*i.e., Canis urostictus* Mivart) were the descendants of these small canids of the John Day. This may very well have been the case, but the long hiatus, during which nothing is known of this phylum, makes positive statements inadvisable. That all South American Fissipedia are the more or less modified descendants of migrants from North America is unquestionably true, but concerning the details of the migration and the steps of evolutionary change, practically nothing is known, for the Pliocene of the southern continent has, as yet, yielded singularly little that bears upon this problem.

Concerning the origin of the White River dogs, more is known, for they seem to be directly derivable from upper Eocene genera of the Uinta stage. The lower Oligocene of the Duchesne River stage is still a problem, so far as its Fissipedia are concerned, but the Uinta *Prodaphoenus* and *Procynodictis*, though incompletely known, have every appearance of being ancestral to the White River genera and, if so, they demonstrate that the two phyla had already become distinct and separate in the upper Eocene. It is not yet practicable to decide whether these Uinta dogs were migrants from the Old World, or whether they were of indigenous origin. Assuming the latter alternative, the Bridger family Uintacyonidae must have been the middle Eocene ancestors sought. It has hitherto been customary to include the Uintacyonidae in the Creodonta but, as was first suggested by the late Professor Matthew, it is more consistent to regard them as the most ancient and primitive of the Fissipedia. It is quite possible that all of the Carnivora were derived from this single family, for the order is a very homogenous and natural group and must have had a common ancestry in the Eocene, but we have still to discover the times and places where the different families arose. That the Uinta genera were directly ancestral to the White River dogs, is altogether probable, but whether those Uinta forms, in turn, were immigrants from Asia, or arose independently in North America, remains to be determined.

In the description of the skeleton, as a whole, and of the various parts, the frequency with which attention is called to viverrine resemblances is very striking. These resemblances are not found in the skull and teeth, which are characteristically cynoid and, in particular, the base of the cranium and its foramina are not in the least viverrine. It is a parallel case with *Daphoenus*, the skeleton of which displays so many feline characteristics, though teeth and skull are not in the least cat-like. Especially does the skeleton of this larger of the White River dogs resemble that of the primitive sabre-tooth *Dinictis*, which also has not a few canine resemblances in the base of the cranium and the foramina. Obviously, any explanation that will account for the feline features of *Daphoenus* will apply equally well to the viverrine characteristics of *Pseudocynodictis*. That those characteristics are viverrine rather than feline is, in all probability, due to the small size of this genus and of most civets.

The comparison might be extended to the weasels, or Mustelidae, which are the subject of the following section of this paper. As will there be shown, the few and rare representatives of this family in the White River, differ very little from the dogs, so far as teeth and skull are concerned. Nothing is known of the skeleton in the Oligocene mustelines, but those of the lower Miocene have cat-viverrine-like skeletons, with long hind legs, powerful loins and long, heavy tail. These facts all lead to the conclusion that *Daphoenus* was not closely related to the cats, nor *Pseudocynodictis* to the civets, but that all these Oligocene fissipeds retain many primitive characters which were common to the early members of nearly all the families, together with specializations which distinguished the families from one another. Subsequently, the specializations became the conspicuous features, as, for example, the transformation of the dogs into pre-eminently cursorial types.

Species. Only a single well-defined species of this genus, *P. gregarius* (Cope), has been named. A supposed second one, *P. lippincottianus* (Cope) has been described, but its status is very doubtful.

Pseudocynodictis gregarius (Cope)

(Pls. XI, Figs. 10, 10a, XIII, Figs. 1–4)

Amphicyon gracilis Leidy (*nec.* Pomel), Proc. Acad. Nat. Sci. Phila., 1856, p. 90.
Amphicyon angustidens Marsh, Amer. Journ. Sci., 3rd ser., II, p. 124.
Canis gregarius Cope, Palaeont. Bull. No. 16, p. 3.
? *Canis lippincottianus* Cope, Synopsis of Vertebr. collected in Colorado in 1873, p. 9.
Galecynus gregarius Cope, Tertiary Vertebrata, p. 916 (1884).
Cynodictis gregarius Scott, Trans. Amer. Phil. Soc., N.S., XIX, p. 400 (1898).

This species is one of the commonest of White River animals and is more frequently met with than are any of the other contemporary carnivores. Despite this abundance of individuals, well-preserved fossils are rare and these are almost always skulls. Only two approximately complete skeletons have been reported, but this is not surprising, when the small size and delicate fragility of the bones are borne in mind.

The *Canis lippincottianus* of Cope is probably merely a large individual of this species, and Cope himself admitted the possibility of that determination. He describes it as having "dimensions half as large again as in *C. gregarius*," but adds: "Unfortunately there is not enough material in my hands to render it clear whether the specimens represent a distinct species, or a large variety of the *C. gregarius.*" In the John Day, there was a greater ramification and four species have been named, not all of them, perhaps tenable.

The species name, *P. gregarius*, requires a word of explanation: Leidy originally named this species *Amphicyon gracilis*, but that term was preoccupied by Pomel's *A. gracilis* and, therefore, Cope substituted *gregarius*, which he first referred to *Canis* and subsequently to the European *Galecynus* Owen. Scott ('98) adopted the term *Cynodictis*, also European,

MEASUREMENTS

	No. 16,381	No. 10,493	No. 10,513	No. 10,939	No. 11,012	No. 11,382	No. 11,432
	mm.	mm.	mm.	mm.	mm.	mm.	mm.
Upper dentition, length i1 to m2				44.0	44.0	43.5	43.5
Upper canine, ant.-post. diam. ..				5.0	5.0	5.0	4.5
Upper canine, transverse diam...				3.5	3.0		3.0
Upper premolar series, length...				25.0		23.0	25.0
Upper molar series, length.....		10.0	10.0	11.0	10.0	10.0	10.0
P1 length (ant.-post.).........				3.5	3.0	3.0	3.0
P2 length...................		5.0		4.5	? 4.0	4.5	3.0
P3 length...................		5.5		5.0			5.5
P4 length...................		6.0		? 4.0		5.5	5.0
P4 width....................		10.0	9.0	9.0	9.5	8.5	9.0
M1 length..................		6.5	7.0	6.0	6.0	6.0	6.0
M1 width...................		9.0		8.0	8.0		
M2 length..................		3.5	3.0	4.0	4.0	3.0	3.0
M2 width...................		6.0		6.0	5.5	4.0	5.0
Lower premolar series, length ...					21.0	19.0	
Lower molar series, length		17.0	16.0		17.0	15.0	
P1 length...................					3.0	3.0	
P2 length...................				5.0	5.0	5.0	
P3 length...................		5.5		5.0	6.0	5.0	
P4 length...................		6.5	7.0	6.0	6.5	6.0	
M1 length...................		10.0	9.5	9.5	10.0	9.0	
M2 length...................		5.0	5.0	5.0	5.0	4.5	
M2 width...................		3.0		3.0	3.5		
Skull, length (fr. occ. condyles).				92.0	92.0	86.0	89.0
Cranium, length (cond. to pre-orb. border)...............		62.0	? 62.0	64.0	64.0	59.0	63.0
Face, pre-orbital length.......				32.0	30.0	30.0	28.0
Occiput, breadth over mast. proc.	35.0	33.0	34.0	34.0	38.0	32.0	33.0
Brain-case, greatest width......	33.0	31.0	32.0	32.0	35.0	33.0	33.0
Skull, width over zygom. arches.		52.0				55.0	
Zygomatic arch, length........		42.0	43.0	43.0	43.0	42.0	44.0
Face, width at p4.............		26.0	26.0	26.0		30.0	25.0
Face, width at canine.........				16.0	17.0	18.0	15.0
Mandible, length fr. condyle....				63.0		60.0	
Mandible, depth at m1........		9.0	11.0	11.0	11.0	10.0	
Mandible, depth at p2........				10.0	8.0	7.0	
Mandible, height of coronoid...		27.0	29.0	? 27.0		29.0	
Mandible, height of condyle....		14.0				13.0	

for this genus and the change was very widely followed, until Schlosser showed that it was untenable and that a new generic term must be coined for the small dogs of the North American Oligocene.

The subjoined table gives the dimensions of the teeth of several individuals in the Princeton University Museum; it will be observed that there is very little variation in size among them, a fact which, so far as it goes, favours the acceptance of P. lippincottianus as a distinct species.

Measurements of the vertebrae have to be made from several individuals, not so much to determine individual variations, as to obtain the necessary data, the vertebrae being so scattered among different specimens. The skeletons in the American Museum and in the South Dakota School of Mines, are so largely in the matrix as to make measuring very difficult.

MEASUREMENTS

	No. 10,493	No. 11,012	No. 11,381	No. 11,382	No. 11,432
	mm.	mm.	mm.	mm.	mm.
Atlas, length..............................	16.0				
Atlas, breadth.............................	34.0				
Axis, length (excl. of odontoid).............	19.0		20.0		
Axis, breadth of ant. face..................	13.0		13.5		
Third cervical, length......................	11.0	13.0	12.0		13.0
Fourth cervical, length.....................					14.0
Fifth cervical, length.......................					13.0
Sixth cervical, length......................			13.0		12.0
Seventh cervical, length....................			11.0		10.0
Anterior dorsal, length.....................		8.0	9.0		8.5
Last dorsal, length.........................	12.0	12.0	13.0		
First lumbar, length........................			15.0	13.0	
Second lumbar, length......................			17.0	14.5	
Fifth lumbar, length........................		16.0	18.0	16.0	
Sixth lumbar, length........................		15.0	17.0	14.0	
Seventh lumbar, length.....................			13.0	13.0	12.0
Sacrum, length.............................		24.0	26.0		
First sacral, width over pleurap.............		24.0	24.0		
Third sacral, width over trans. proc..........		21.0			
First caudal, length........................		7.0	8.0	? 10.0	
First caudal, width over trans. proc..........		21.0	26.0		
Median caudal, width of ant. face...........		5.0			

Dimensions of the fore leg must also be obtained from several individuals and in these we note a greater range of variation in size. It will be observed that in the following table No. 11,381 is considerably larger than any of the others.

For the manus there are but two available individuals in which substantially all the parts, phalanges excepted, have been collected. No phalanges, sufficiently well-preserved to make measurements worth while, are preserved in connection with any of the individuals which have supplied the materials for the foregoing tables of dimensions.

MEASUREMENTS

	No. 10,493	No. 11,012	No. 11,381	No. 11,382	No. 11,492
	mm.	mm.	mm.	mm.	mm.
Scapula, length............................		54.0			
Scapula, greatest width...................		? 49.0			
Scapula, width of neck....................			13.0		
Scapula, ant.-post. diam. of glen. cavity......			12.0	9.5	9.5
Scapula, transverse diam. of glen. cavity.....			8.0	7.0	7.0
Humerus, length..........................		75.0		70.0	70.0
Humerus, ant.-post. diam. prox. end........	12.0	15.0	19.0	15.0	15.0
Humerus, transverse diam. prox. end........		14.0	16.0	13.0	12.5
Humerus, breadth of dist. end..............		16.0	20.0	15.0	
Humerus, breadth of trochlea..............		12.0	14.5	11.0	
Radius, length............................	57.0	61.0			
Radius, ant.-post. diam. prox. end..........	5.0	6.0	7.0	5.0	5.0
Radius, transverse diam. prox. end..........	7.0	7.0	9.0	7.0	7.0
Radius, breadth of dist. end...............	12.0	13.0	13.0		9.0
Radius, breadth of carpal surface..........	5.5	6.0	7.0		5.5
Ulna, length..,..........................		72.0			
Ulna, length of olecranon...................		7.0	10.0	9.5	9.0
Ulna, thickness of olecranon...............			10.0	8.0	8.0

MEASUREMENTS

	No. 10,493	No. 11,012
Carpus, height in median line.....................	mm.	6.0 mm.
Carpus, breadth................................		11.0
Metacarpal I, length............................		12.0
Metacarpal I, width of prox. end..................	3.5	4.0
Metacarpal I, width of dist. end..................		3.0
Metacarpal II, width of prox. end.................		3.5
Metacarpal III, length...........................	22.0	21.5
Metacarpal III, width of prox. end................	4.0	3.5
Metacarpal III, width of dist. end................	5.0	4.5
Metacarpal IV, width of prox. end.................	4.0	3.5
Metacarpal V, length............................	17.0	16.0
Metacarpal V, width of prox. end.................	4.0	4.0
Metacarpal V, width of dist. end.................	4.0	5.0

A different set of individuals must be employed for taking the dimensions of the hind limb.

In the material before us, the hind foot is the most incompletely represented part of the skeleton. The New York skeleton, at the time of writing, is not available.

MEASUREMENTS

	No. 11,012	No. 11,381	No. 11,382	No. 11,432
	mm.	mm.	mm.	mm.
Pelvis, length............................	? 64.0			
Pelvis, breadth at acetabulum..............	36.0	37.0		
Ilium, length from acetabulum.............	? 33.0	37.0		
Ilium, breadth of peduncle.................	11.0	10.0	9.0	
Ilium, breadth of ant. plate................			13.0	
Ischium, length from acetabulum...........	27.0		26.0	
Femur, length............................		93.0	85.0	86.0
Femur, breadth of prox. end...............	17.0	20.0	15.0	16.0
Femur, breadth of dist. end...............	16.0	17.0	14.0	14.0
Tibia, length.............................	89.0	99.0		
Tibia, breadth of prox. end...............	15.0	18.0	14.0	14.0
Tibia, breadth of dist. end................	11.0	12.0	9.0	9.0
Fibula, thickness of prox. end.............	7.0			
Fibula, thickness of dist. end.............		9.0		

	No. 10,493	No. 11,012	No. 11,381
	mm.	mm.	mm.
Tarsus, height, excl. calcan............................	21.0		
Calcaneum, length......................................	19.5	20.0	
Calcaneum, length of tuber............................	12.0	12.0	
Calcaneum, dorso-plant. diam..........................	7.0	8.0	
Astragalus, length.....................................	13.0	13.0	14.0
Astragalus, width of trochlea...........................	5.0	5.5	6.0
Astragalus, length of neck.............................	6.0	6.0	6.0
Astragalus, width of head..............................	7.0	7.0	8.0
Navicular, prox.-dist. height...........................	3.0		
Navicular, width......................................	6.0		
Ecto-cuneiform, height................................	4.5		
Ecto-cuneiform, width dist. end........................	4.5		
Metatarsal I, width prox. end..........................	4.5		
Metatarsal II, width prox. end.........................	3.0	3.0	
Metatarsal III, width prox. end........................	5.0	3.0	
Metatarsal III, width dist. end.........................		5.0	
Metatarsal IV, width prox. end.........................	3.5		

Parictis Scott

(Pl. XIV)

Parietis Scott (*errore typog.*), Amer. Natur., XXVII, p. 658, 1893.
Parictis Lydekker, Zool. Record for 1893, XXX, Mamm., p. 29.
Parictis E. R. Hall, Journ. Mammal., 12, p. 156, 1931.

This genus, originally described from the John Day formation and erroneously referred to the Mustelidae, has lately been found in the Chadron substage of the White River. Hall was the first to show that the genus, while distinct from *Pseudocynodictis*, is properly referable to the Canidae. So far, only lower jaws have been found and the White River specimen is in the collection of the South Dakota School of Mines at Rapid City.

In *Parictis* the lower dental formula is presumably the same as in *Pseudocynodictis*, the only uncertainty being as to the number of incisors, but there is no reason to think that they were less than three: i?$\overline{3}$, c$\overline{1}$, p$\overline{4}$, m$\overline{3}$. The premolars are set much more closely together than in the genus last named, each tooth slightly overlapping the one behind it; they are of similar simple, compressed-conical and trenchant shape, but markedly thicker transversely, recalling the form seen in the European genus *Cynodon*, to which *Parictis* would seem to be related. If so, the latter must have been a migrant from the Old World. The blade of the sectorial (m$\overline{1}$) is relatively smaller and lower, the heel larger and more basin-like, because of the higher internal cusp than in *Pseudocynodictis* and the tubercular molars, (m$\overline{2}$ and $\overline{3}$) are less reduced.

Dr. Clark suggests that in *Parictis* we have the earliest recognizable member of the series which led through *Phlaocyon*, of the lower Miocene to the Procyonidae, which are not known as such before the Pliocene.

Parictis dakotensis sp. nov. (Clark, Ms.)

(Pl. XIV, Figs. 1, 1a)

Dr. Clark has kindly supplied the following notes from the Ms. of his forthcoming paper on the Chadron Formation.

"*Type* right *ramus mandibuli* with p$\overline{2}$, $\overline{3}$, $\overline{4}$, m$\overline{1}$ and $\overline{2}$, and alveoli for $\overline{c}$, p$\overline{1}$ and m$\overline{3}$, belonging to the museum of the South Dakota State School of Mines, Rapid City.

"*Specific characters:* tooth-row somewhat longer than that of *P. primaevus:* p$\overline{3}$ larger than p$\overline{2}$ and much expanded posteriorly; in *P. primaevus* p$\overline{3}$ is slightly shorter than p$\overline{2}$ and is not expanded posteriorly. Jaw very much deeper and more robust, actually and proportionally, than that of *P. primaevus.*

"*Parictis dakotensis* is a rather generalized form, with molars and premolars about equally developed, *Phlaocyon*, with almost exactly the same cuspidation, has the premolars reduced and the molars enlarged. *Parictis dakotensis*, or some species yet unknown and very closely allied to it, may be regarded as the Chadron ancestor of *Phlaocyon* and hence (*fide* Wortmann and Matthew) of all the raccoons.

"Wortmann and Matthew suggested *Cynodictis* [*i.e.*, *Pseudocynodictis*] as a possible ancestor of *Phlaocyon;* Teilhard de Chardin in 1915 believed that *Pachycynodon* was possibly the ancestral form. *P. dakotensis* agrees with the Princeton Museum specimens of *Cynodon* and with Teilhard de Chardin's figures of *Cynodon* and *Pachycynodon* in the general tooth cuspidation, in the tendency toward subquadrangular premolars and toward a cingulum on the premolars.

"The teeth of *Parictis* are broader, heavier and blunter than in most of the species of the European genera and the jaw is more massive. *Pseudocynodictis* has sharp, high, blade-like premolars, m$\overline{1}$ much larger than p$\overline{4}$ and is, in general, of a lighter, more delicate build than is *Parictis*. Such evidence as the lower jaws can offer, therefore, indicates a relationship between *Parictis* and *Cynodon-Pachycynodon*, rather than between *Parictis* and *Pseudocynodictis*."

Horizon: Chadron.

Family 2. MUSTELIDAE

One of the strongest contrasts between the Oligocene mammalian fauna of North America and that of Europe is in the far greater abundance and variety of this family in the Old World than in the New. In the White River beds mustelines are very rare as fossils and of very limited variety and they have not yet become clearly distinguished from the dogs, from the primitive representatives of which they were probably descended.

Subfamily MUSTELINAE

An interesting discovery by Dr. Clark in the collecting season of 1934 is that of a new and undescribed weasel and the following description is taken from his manuscript, with certain abbreviations.

Mustelavus priscus gen. et sp. nov. (Clark Ms.)

(Pl. XIV, Figs. 2, 2a)

"*Type*, somewhat crushed skull, with lower jaws in occlusion, Princeton Museum, No. 13,776.

"*Horizon:* upper member, Chadron formation, lower Oligocene.

"*Generic and specific characters:* Skull long and slender; very faint parietal crests, close to middle line; basi-cranium long, palate rather broad relative to entire skull; bullae moderately inflated. Dentition $\frac{3-1-4-2}{?-1-4-2}$; i$\underline{1}$ smallest, i$\underline{3}$ largest, all very slender; $\underline{c}$ slender, slightly curved, pointing anteriorly; p$\underline{1}$ single-rooted; p$\underline{2}$ double-rooted; p$\underline{3}$ simple, sub-conical, strongly compressed laterally; with a well defined posterior basal cusplet; p$\underline{4}$ rather small, protocone very little higher than that of p$\underline{3}$, deuterocone not extended very far inward. M$\underline{1}$ moderately high-cusped, parastyle prominent, set external to paracone; m$\underline{2}$ alveolus small.

"Mandible long, delicate, laterally compressed; condyle in line with tooth-row and internal end directed antero-ventrally; c̄ short, sharp, conical; p$\overline{1}$ small, single-rooted, cusp directed anteriorly, heel vertically; p$\overline{2}$ simple, two-rooted, with anterior and posterior cingulum and rather long-heel; p$\overline{3}$ like p$\overline{2}$, but slightly larger; p$\overline{4}$ slightly larger than p$\overline{3}$, with small, but distinct deuteroconid. M$\overline{1}$ trigonid distinctly higher than p$\overline{4}$, protoconid overtopping the equal para- and metaconids and set slightly anterior to the metaconid; talonid low, slightly basined. M$\overline{2}$ small, sharply cusped, having an anterior and a posterior basin, separated by a low elevation connecting protoconid and metaconid.

"Comparison of the type lower jaw with the specimens of the European *Plesictis* and with Teilhard's description and figures shows no good basis for a generic separation. In fact, the resemblance to *P. pygmaeus* is so close, that it is almost impossible to find ground for a specific distinction in the lower jaws. Comparison of the skull with Filhol's figures of *Plesictis* reveals one sharp difference, retention of m$\underline{2}$ in *Mustelavus* and its absence in *Plesictis*. Also, the parietal crests of the latter are more strongly developed than those of the American genus. Otherwise, the two are extremely similar in dentition, in skull configuration and in basi-cranial anatomy.

"In keeping with its stratigraphic position, *Mustelavus* exhibits many primitive and generalized characters. Retention of m$\underline{2}$, parastyle of m$\underline{1}$ relatively small and internal basin not extending anterior to protocone; metaconid of m$\overline{1}$ as large as paraconid; m$\overline{2}$ long antero-posteriorly and sub-rectangular, are all characters which may be regarded as primitive within the subfamily. The recent genera have all lost m$\underline{2}$, have the parastyle of m$\underline{1}$ large and the internal basin encircling the protocone, metaconid of m$\overline{1}$ very small to absent and m$\overline{2}$ sub-rectangular to round, although with essentially the same cuspidation as m$\overline{2}$ of *Mustelavus*. The same dental specializations are present in *Gulo* as in the Mustelinae.

"The dentition of *Oligobunis* is, except for its much greater size and massiveness, extremely like that of *Mustelavus*. Reduction of the metaconid of m$\overline{1}$ to a small tubercle on the inner side of the protoconid is the only real advance.

"*Mustelavus* is, therefore, representative of the ancestral stock from which *Oligobunis*, the recent Mustelinae and Guloninae were derived. *Plesictis*, lacking m2, is much less satisfactory as an ancestral form, inasmuch as m$\underline{2}$ is present in the American Miocene mustelines (*Oligobunis*, *Paroligobunis*, *Brachypsalis*). *Bunaelurus*, with the protocone of m$\underline{1}$ a simple transverse blade, lacking any encircling basin, the metaconid of m$\overline{1}$ absent, m$\overline{2}$ and the heel of m$\overline{1}$ trenchant rather than basined, is markedly different from the other American Mustelinae and cannot be regarded as ancestral to them."

Bunaelurus Cope

(Pl. XIV)

Bunaelurus Cope, Ann. Rept. U. S. Geolog. & Geogr. Surv. Terrs. for 1873, p. 507 (1874); Matthew, W. D., Bull. Amer. Mus. Nat. Hist., XVI, p. 137 (1902).

The late Dr. Matthew described and figured (*loc. cit.*) what was for a long time, the only known skull of this rare animal and his account may, with advantage, be reproduced here in abbreviated form. Another fine skull is in the Princeton collection.

"The characters of the skull confirm the views expressed by Cope and Schlosser as to the position of the genus. It is *Palaeogale* with a minute second molar still retained. It belongs to the primitive division of the Mustelinae, with triangular first molar, no posterior flange on the protocone. The carnassial is primitive in character, somewhat resembling that of *Cynodictis* [*i.e.*, *Pseudocynodictis*] *gregarius*, the protocone very large, the shear more oblique than in modern Mustelinae, less so than in *Cynodictis*, the fissure between protocone and postero-external blade still quite well marked. There is a small antero-internal basal cusp and a less marked antero-external one. The second and third premolars are of moderate size without trittocones [*sic*] much higher than in *Mustela*, higher and proportionately larger than in *Putorius*. The first premolar is a single-rooted tooth of small size; first and second premolars are spaced. Alveoli of canines, of moderate size, are preserved.

"The bullae are of primitive character, inflated, short and prominent, instead of flattened and elongated as in *Mustela* and *Putorius*. The palate extends backward only to a point opposite the anterior edge of the first molar, while in modern Mustelinae it extends considerably behind those teeth. The shorter bullae leave a much larger surface of the sphenoids and occipitals exposed; the short, stout paroccipital process is entirely free of the bulla.

"The occipital and sagittal crests have the same outlines as in *Putorius ermineus*, but the posterior lobes of the brain are separated from the cerebral lobes by a strongly marked depression; the arches are much heavier, muzzle much longer, resembling that of *Mustela* more nearly, but flatter, longer, more slender toward the tip; infraorbital foramen smaller, postorbital process of the frontal less prominent. Postorbital constriction much more narrow than in *M. americana*, somewhat more than in *P. ermineus*. Size slightly greater than that of the weasel.

"*Bunaelurus* is one of the primitive group of Mustelinae found chiefly in the Oligocene of Europe. It belongs to the Putoriine section, which more nearly approaches the Felidae (through *Proaelurus*) in dental reduction (the typical Musteline section more nearly approaching *Cynodictis* and the Viverridae) but shows little indication of the shortening of the face characteristic of the modern *Putorius*."

"The White River skull under discussion belongs unquestionably to the Putoriine group and with the primitive members thereof."

Bunaelurus lagophagus Cope

(Pl. XIV, Figs. 3–3*a*)

Bunaelurus lagophagus Cope, *op. cit.*, p. 507.

Bunaelurus infelix Matthew

Bunaelurus infelix Matthew, Bull. Amer. Mus. Nat. Hist., XIX, p. 210 (1903).

These species are "with difficulty distinguishable," but that they are probably distinct, nevertheless, is shown by the fact that *B. lagophagus* is from the Brulé substage and *B. infelix* from the Chadron. Matthew gives the following dimensions of the latter.

MEASUREMENTS

P4̄–m2̄	10.2 mm.	M1̄, longit.	5.9 mm
P4̄, longit.	4.1	M1̄, transverse	2.3
P4̄, transverse	1.9	Depth of jaw beneath m1̄	6.0

The White River mustelines are of Old World type and were obviously immigrants; no possible ancestors of them are known from the American Eocene.

Superfamily AELUROIDEA

Family 3. FELIDAE

The proper subdivision of the cats and their relations to other families of the Fissipedia are difficult problems, concerning which there are still wide differences of opinion. As a tentative arrangement may be suggested the formation of three subfamilies: (1) Felinae, true cats, which include all living members of the family, the other groups being extinct. No true cat has been found in North America before the Pliocene and they made a late appearance in Europe also (middle and upper Miocene).

(2) Machairodontinae, or sabre-tooth cats, in which the upper canines have been converted into great, recurved, laniary tusks, with serrate edges, which in some genera (*e.g.*, *Eusmilus*, *Smilodon*) attain astonishing proportions. This subfamily, though very like the Felinae in general appearance, differs from them markedly in the structure of the cranial basis, the glenoid and mastoid regions and in the form of the mandible. The American history of this subfamily, so far as is yet known, begins in the Chadron substage, runs through the White River and John Day and after a long hiatus in the Miocene, reappears in the Pliocene and culminates in the Quaternary of North and South America. (3) Nimravinae, or "false sabre-tooths," as Cope, their discoverer, called them; these are, in a sense, intermediate between the other two subfamilies, though most like the machairodonts, in which they are usually included. The uppermost White River has yielded the typical genus *Nimravus* and the group continues through the John Day and lowest Miocene

and after a long hiatus reappears through the middle and upper Miocene and lower Pliocene and then dies out.*

Subfamily MACHAIRODONTINAE

These are the sabre-tooth cats, so called because the upper canines are great scimitar-like tusks, very broad antero-posteriorly, thin transversely, with serrate edges, which in some genera (*e.g.*, *Eusmilus*) reach exaggerated size. In proportion as the upper canines are enlarged, the lower ones are reduced. Auditory bulla mostly simple, or unossified, not touching paroccipital or mastoid processes. Cranial foramina more canine than feline. Mastoid and root of zygomatic process forming conspicuous, massive, dependent pedicles unlike those of any other fissipeds, the size of which increases with that of the sabres. Lower jaw flanged anteriorly for protection of the tusks.

Dinictis Leidy

(Pls. XV, XVII, XVIII, XXII)

Dinictis Leidy, Proc. Acad. Nat. Sci. Phila., 1854, pp. 127, 156.
Deinictis Leidy, *ibid.*, 1856, p. 91.

This genus is the most primitive known member of the subfamily and has many points of resemblance to the early dogs, such as *Daphoenus*. The following description, teeth and skull excepted, is taken from the remarkably complete skeleton of *D. squalidens* in the Museum of Comparative Zoology, Harvard University; supplemented by material in several other museums, notably those of Rapid City, New York, and Princeton.

DENTITION

The dental formula is: $i\frac{3}{3}$, $c\frac{1}{1}$, $p\frac{3}{3}$, $m\frac{1}{2}$, the formula which recurs in most of the Mustelidae and also in *Aelurogale*.

Upper Teeth. The *incisors* are, in general, cat-like and form a straight, transverse row, separated from the canines by considerable diastemata. The external incisor (i3) is the largest of the series and has a long, acutely conical and somewhat recurved crown and, except for smaller size, is very like i3 in *Drepanodon*. As in that genus, there is no cingulum such as occurs on that tooth in modern cats. The second incisor (i2) is much smaller than the external one and has a simple, pointed crown, without cingulum; the median incisor (i1) is still smaller and more compressed laterally than the second.

The *canine* is a long, thin, recurved and sabre-like tusk, with finely serrate borders, but the tusk is shorter, thicker and less compressed than in *Drepanodon*, to say nothing of *Eusmilus*, in which the sabre attains exaggerated proportions.

The *premolars* are three in number, the first one (p1) having been suppressed; the most anterior cheek-tooth is thus p2 and is exceedingly small, but is implanted by two roots; it has a compressed-conical crown, with sharp and finely serrate posterior edge and very minute posterior basal cusp. P3 is large and well-developed, and is much higher and more compressed laterally than in the true cats, and the large posterior cusp is single, not double as in the latter. The sectorial, p4, is more canine in form than either feline, or vi-

* This phylogeny and classification are proposed by the senior author. Another arrangement, somewhat similar to that of the late W. D. Matthew, which interprets certain cranial structures differently, is preferred by the junior author.

verrine and differs from p$\underline{3}$ only in its greater size, the larger posterior blade and the presence of an inner cusp. This tooth resembles the sectorial of *Daphoenus*. Of the anterior basal cusp, which is so characteristic of nearly all cats and appears in *Cryptoprocta* and certain other viverrines, there is no sign.

The single upper molar (m$\underline{1}$) of *Dinictis* is very much better developed than in the true cats, in *Cryptoprocta*, or *Drepanodon*, but is much more reduced than in most of the Viverridae; it is plainly visible in a side view of the skull, for it is not concealed by the sectorial, as it is in the felines. The crown shows clearly its derivation from the tritubercular pattern, but its antero-posterior diameter is greatly reduced, which has brought the two external cusps into nearly the same transverse line, while the inner cusp has been extended in the same direction. The tooth is inserted by three roots.

Lower Teeth. The *incisors*, as in the cats generally, form a straight, transverse row and the root of the middle one (i$\bar{2}$) is not, as it is in the other fissiped families, pushed back out of line with the other two. The outer incisor (i$\bar{3}$) is the largest of the series, as in the upper jaw, and they all have simply conical crowns. In the series of sabre-tooth cats there is an inverse relation in size between the upper and the lower *canines*. In *Dinictis*, in which the scimetar tusk is less elongate than in the more specialized genera, such as *Drepanodon* and *Eusmilus*, the lower canine is much larger than in the genera named and has not become, in appearance, one of the incisors, but is manifestly a canine, with high, sharp-pointed and slightly recurved crown.

The most anterior premolar (p$\bar{2}$) is very small, though it has two roots, and the crown is a simple compressed cone, without accessory cusps. P$\bar{3}$ is relatively larger, higher, more acutely pointed and more compressed than in the true cats; it is composed of a principal compressed cone, with basal cusps before and behind. The last premolar (p$\bar{4}$) is almost exactly like p$\bar{3}$, except that it is somewhat larger and the accessory basal cusps are more conspicuous. Both of these teeth differ from the corresponding premolars of *Felis* not only in their relatively larger size and smaller transverse diameter, but also in the absence of the posterior cingulum.

The sectorial *molar* (m$\bar{1}$) is very feline in character, but with obvious signs of its derivation from the tuberculo-sectorial type exemplified in the Creodonta and in the primitive fissiped family of the Uintacyonidae. It is essentially a bicuspid, trenchant blade, of which the anterior cusp is like that of the true felines, but the posterior one is higher and less flattened and is more distinctly angulate between the lateral and posterior surfaces. In many individuals there is a small but perfectly distinct postero-internal cusp (metaconid), thus completing the primitive triangle of the tuberculo-sectorial tooth, and in these individuals the tooth is almost a reproduction of the lower sectorial of *Proaelurus*, as figured by Filhol. In *Dinictis*, however, this cusp was evidently on the point of disappearance for, in some specimens it is present only on one side of the jaw and in others it is hardly visible at all. Another remnant of the tuberculo-sectorial tooth is the small heel, or talon, which has a sharp cutting edge and no trace of accessory tubercles. The tubercular molar (m$\bar{2}$) is very much reduced and has a small oval crown. The root, though single, was evidently formed by the coalescence of two roots, as is shown by the groove which runs down the inner side of the root and by the partial division of the alveolus. This tooth is much less reduced than in *Proaelurus*.

The dentition of *Dinictis*, while clearly indicative of relationship with the cats, retains several persistent primitive characteristics, such as the greater number of teeth, the large size of the single upper molar and the presence of a second lower molar, the dog-like upper sectorial with no antero-external basal cusp, the lower sectorial with internal cusp and vestigial talon. In these respects, the genus departs from the feline and approximates the canine type.

SKULL

As in primitive fissipeds generally, the cranium is long and narrow and is deeply constricted some distance behind the orbits; this postorbital constriction marks the anterior boundary of the cerebral hemispheres and is farther behind the orbits than in modern cats or viverrines, occupying the same relative position as in such primitive dogs as *Daphoenus*. The posterior part of the cranium, or that portion of it which is behind the mastoid processes, is very elongate, as in *Daphoenus*, *Pseudocynodictis*, the viverrines and most creodonts. Notwithstanding the very long cranium, the cerebral fossa is relatively short, while the cavities for the hind-brain and the olfactory lobes are correspondingly large.

In the different species of the genus there are two types of skull, which have a very different appearance, though, in all probability, the difference was not really important in the living animals. In the larger species, including the type, *D. felina*, the highest point of the skull is just behind the orbits and from this point the upper contour of the brain-case slopes steeply downward and backward to the occipital crest, while that of the face slopes, somewhat less steeply, downward and forward from the same point, where the two meet in an open V. At the point of meeting the skull has a disproportionate vertical height, while the muzzle is low and the occiput still lower. Notwithstanding the curiously sloping brain-case, the cranio-facial axis is straight.

The second type of skull, exemplified in the smaller species, such as *D. squalidens*, of the White River, and *D. cyclops*, of the John Day, the brain-case has no such slope downward and backward and therefore a more normal appearance. There are other differences of greater morphological significance, which are associated with these two types of skull, as will be pointed out in a subsequent section.

The occiput is low and very broad at the base, narrowing rapidly upward; it is exceedingly convex, in correspondence with the very anterior position of the mastoids and, consequently, much of the occipital surface is visible, when the skull is seen from the side. Above the foramen magnum there is a strong convexity produced by the prominent vermis of the cerebellum. The paroccipital processes are short, almost vestigial, in fact; they project more or less directly backward and are separated from the auditory bullae by considerable intervals, as in the mustelines. The mastoid processes, which are also removed from the bullae, differ in the two types of skull; in the larger species they are very prominent, rugose and heavy and brought so near to the post-glenoid processes that only narrow passages lead into the auditory openings. The great enlargement of the mastoids, which is carried much farther in *Drepanodon* and reaches an extraordinary development in *Eusmilus*, supplies an increased area for the sterno-mastoid muscles and thus favours the hypothesis of Professor Matthew and of Dr. Merriam, that the machairodonts used their great sabres to strike a stabbing blow, as does a snake. This problem will be considered in connection with *Drepanodon* and *Eusmilus*, in which the sabres attained such an immense size, that they would seem to have defeated their own purpose and to have become useless. Such a conclusion is a *reductio ad absurdum* and could not be admitted, save under the most cogent and irrefragable evidence.   The development of the mastoid process is very much less advanced in *Dinictis squalidens*, in which the second type of skull, with horizontal brain-case occurs.

The cranial base is broad and the glenoid cavities of the two sides are widely separated, which results in a somewhat unusual arrangement of the foramina.   The basioccipital is wide and deeply concave transversely, the sides being turned up into sharp, prominent edges.   The fossa in which the auditory bulla was placed is large and must have been filled by an extensive, inflated structure, which probably was not ossified, for no skulls, with ossified bullae attached, have yet been reported.   It is, of course, possible that the tympanics were so loosely attached, that they were almost invariably lost in fossilization, as is true of the baculum and the clavicles, for instance, and, for the present, the question must remain undecided.

The *parietals* are very large, relatively larger than in the true cats and form nearly all the roof and much of the sides of the cerebral fossa.   The sagittal crest is very long, extending past the postorbital constriction upon the frontals.   The *squamosal* is rather low vertically, but elongate antero-posteriorly.   The root of the zygomatic process projects downward below the basicranial axis much farther than in the true cats and civets, but much less than in *Drepanodon*, which, in turn, is surpassed by *Eusmilus* and *Smilodon*, in all of which it forms a peculiar and highly characteristic pedicle.   This develops *pari passu* with the enlargement of the mastoid, which is in approximate proportion to the size of the canine sabres.   The formation of these parallel pedicles is one of the most characteristic features of the machairodont subfamily, as distinguished from the other two subdivisions of the Felidae.   To a certain extent, *Dinictis* is thus intermediate between the more specialized Machairodontinae and the Nimravinae.   The glenoid cavity is cat-like owing to the prominence of the preglenoid ridge.   The zygomatic arch is relatively long, compressed, but stout, and it does not curve out so far from the side of the skull as it does in modern cats; the postorbital process of the jugal is low, leaving the orbit widely open behind.

The limits of the *sphenoid* bones are rarely determinable, sometimes because of the closing of the sutures, sometimes because the skull is much cracked, though without distortion.

The *frontals*, which are relatively smaller than in the Felinae, are, in the larger species, steeply inclined downward and forward, in the smaller species more nearly horizontal. The sagittal crest extends over upon these bones in advance of the postorbital constriction where it divides into the temporal ridges.   The postorbital processes are shorter than in modern felines and the orbits correspondingly open.   Anteriorly, the frontals are deeply notched to receive the ends of the nasals, but the fronto-nasal processes are very short and more obtuse than in modern cats, the maxillary and nasal edges meeting at right angles.   The *lachrymal* is small and is not extended upon the face and just within the edge of the orbit it is pierced by two foramina.

The *nasals* are longer than in the true cats, but otherwise very much like them.   The *premaxillaries* have thick and well-developed alveolar portions, which form a nearly straight line, as the bones are abruptly truncated in front, and project but little in advance of the canines; the ascending ramus is very long and nearly vertical and is far removed from the frontal.   The *maxillary* has a high and antero-posteriorly narrow pre-orbital portion, which joins the frontal by a short, straight suture. The suborbital part of the maxilla exceeds the preorbital portion in length more than it does in the true cats. In spite of the transverse thinness of the canine sabre, it produces a more conspicuous prominence on the face than does the thicker fang of *Felis*.

The hard *palate* is very broad behind and, following the oblique tooth-rows, it narrows rapidly forward. The palatine processes of the premaxillaries are unusually large and the spines broad and strong; those of the maxillaries are also extensive. The *palatines* form a considerable share of the bony palate and their anterior border is a straight transverse line; the posterior palatine foramina are very small. The *posterior nares* are very long, because the palatines are not extended behind the teeth. In front, the opening is broad and its anterior edge is formed by two shallow emarginations of the palatines, separated by a median spine. The walls of the canal, formed by the palatines and *pterygoids*, are elongate and constricted in the posterior part, which gives them a characteristic shape. The hamular processes of the pterygoids are very feline both in form and in position and are much farther forward than in the civets, and the whole palatine region is cat-like in character.

The *mandible* is peculiar and is, in most respects, very much like that of *Drepanodon*, but the condyle is raised more above the level of the teeth and the coronoid process is much higher, it is narrower antero-posteriorly, straighter and less recurved than in the felines generally. These differences in shape of the mandible between the two sabre-tooth genera are in manifest correlation with the development of the glenoid and mastoid pedicles. In *Dinictis* the masseteric fossa is deep and extends forward to a point beneath m$\overline{2}$. The horizontal ramus is compressed and rather slender and shallow, its ventral border is nearly straight except that near the anterior end is a small bony flange for the protection of the upper canine; in *Drepanodon* the flange is much larger and in *Eusmilus* it is extremely enlarged in correspondence with the length of the sabres. The chin is slightly concave and meets the lateral surface of the ramus at right angles; the symphysis is short and nearly vertical.

The *foramina* are a curious assemblage, being anything but "aeluroid" in number and position. The incisive foramina are like those of the true cats, but the posterior palatine foramina are small and perforate the maxillaries, not the palatines, and are opposite the anterior edge of p$\underline{3}$. There is an alisphenoid canal, which is not present in the true felines, the posterior opening of which is enclosed in a common groove with the very large *foramen ovale*, separated by a prominent ridge from the eustachian canal and the *foramen lacerum medium*, which occupy the usual position. There is a large and distinct carotid foramen, which is not enclosed in the *foramen lacerum posterius*, from which the condylar foramen is also separate, and there is no bony ridge running mesially from the paroccipital process; the glenoid foramen is very large. No existing carnivore displays such an assemblage of characters of the *basis cranii*, which is perhaps more cynoid than anything else.

VERTEBRAL COLUMN, RIBS AND STERNUM

The vertebral formula is: C 7, D 13, L 7, S 3, Cd 17+. The neck has a somewhat deceptive appearance of length, though it is in fact, considerably shorter than the skull, and the vertebrae are small and weak in comparison with those of a modern cat, such as the Leopard. The *atlas*, so far as it is preserved, is feline in character; the inferior arch is very slender; the neural arch is much broader and its forward extension has converted the atlanteo-diapophysial notches into foramina; there is no indication of a neural spine.   The transverse processes are broken, but it is plain that they extend much farther forward in attachment to the neural arch than in *Felis;* the hinder opening of the vertebrarterial canal has the same position as in the latter, but the anterior opening is farther forward.

The *axis* is also very cat-like in shape, but differs in several details from that of *Felis.* The centrum is narrower and shorter and the anterior cotyles do not project out so far laterally and there is only a faintly marked ventral keel.   The pedicles of the neural arch are narrower from before backward and the post-zygapophyses have a more anterior position.   The neural spine, while having the hatchet-like shape usual among the fissipeds, differs from that seen in *Felis;* anteriorly, the plate is not produced nearly so far in front of the pedicles and, posteriorly it extends much farther behind the post-zygapophyses so as to overhang the fourth vertebra, while in *Felis* it does not reach to the middle of the third. The posterior borders slope upward and backward at an angle of about 45°, whereas in the modern genus they are nearly vertical.   As regards the form of the axis and especially of its neural spine, *Dinictis* and *Daphoenus* are more nearly alike than either one of them is to Recent dogs or cats.

The other *cervicals* are relatively small and weak, with neural spines shorter and all the processes less developed and sturdy, though like those of *Felis* in shape.   On the fifth the transverse process projects more freely from the inferior lamella and on the seventh, as usual, there is no inferior lamella and no canal for the vertebral artery.

The anterior *dorsal vertebrae* have small centra and remarkably short and weak neural spines; in no part of the skeleton is the contrast with the Leopard more striking than in these vertebrae.   The spines slope backward to the tenth, which is the anticlinal, with short erect spine; while the eleventh has no distinct spine; the twelfth and thirteenth dorsals have spines of the lumbar type, inclined forward.   The transverse processes are conspicuous and have facets for the tubercles of the ribs to the eleventh inclusive; the twelfth and thirteenth have no transverse processes.   Zygapophyses change their character on the eleventh dorsal, where the post-zygapophyses are raised high above the anterior pair and take on the lumbar shape, presenting laterally instead of ventrally.   Metapophyses are present in incipient form on the eleventh, becoming progressively higher and more massive on the twelfth and thirteenth.   Anapophyses are also present on the eleventh, twelfth and thirteenth dorsals, increasing in size posteriorly.

The *lumbar vertebrae* are disproportionately larger than the dorsals and make the loins very long, but compared with those of the Leopard, these vertebrae, though elongate, are relatively weak, and in remarkable contrast to those of *Felis.*   The centra are long and slender, especially in the transverse diameter; the neural spines, which incline forward steeply, increase in height to the sixth vertebra, the seventh is much shorter, the shortest of the series in fact.   Metapophyses are very prominent and increase in height posteriorly from the twelfth dorsal to the fourth, fifth and sixth lumbars.   Anapophyses, which are very large and prominent on the last three dorsals, and the first lumbar, diminish rapidly in size backward, becoming vestigial on the fourth and fifth and absent from the last two. The transverse processes are astonishingly short and weak; they are wanting on the first and second, very short on the third, but increase in size to the seventh.   The contrast with the Leopard is very striking, in the breadth and thickness of the centra, the height and width of the neural spines, the length and width, forward and downward curvature of the transverse processes; in all these features the lumbars of *Dinictis* are incomparably weaker and more slender, but they are very long, approximating those of the Leopard, a much larger animal, in actual size and making very elongate loins.

The *sacrum* is long and narrow, consisting, as in most fissipeds, of three vertebrae, of which only the first has pleurapophyses for the support of the pelvis. The neural spines, which are relatively much higher than in *Felis*, remain, as in that genus, widely separated from each other

The tail, which seems to be almost complete, is of only moderate length, much less than in the contemporary sabre-tooth *Drepanodon*, or the White River dogs, such as *Daphoenus* and *Pseudocynodictis*. In this skeleton, the number of *caudal vertebrae* is seventeen but the terminal one and several intermediate ones are evidently missing. All of these vertebrae are conspicuously smaller than the corresponding ones of the Leopard. Neural spines are prominent on the first and second caudals and metapophyses are distinguishable, diminishing in size, back to the seventh. Transverse processes, stout, depressed and inclining outward and backward continue as far as the fifth, behind which they become lateral ridges, with anterior and posterior ends more prominent, but very much less so than in the Leopard. In the middle region of the tail the processes are all much more completely reduced than in the large modern cats (not including *Lynx*, of course) and the centra gradually diminish in length and thickness in quite the usual way.

Ribs. The anterior ribs are very short and stouter than the others, which are slender and rod-like in proportion to the size of the animal; in length, they increase to the eighth and diminish behind that point. Tubercles, for articulation with the transverse processes of the dorsal vertebrae, are large as far back as the eleventh pair of ribs.

The *thorax*, as a whole, is short in proportion to the length of the trunk; actually shorter than the lumbar region; it is also shallow and narrow, as in the "vermiform" mustelines and civets.

The *sternum* is very feline in character; the manubrium has a prominently projecting pair of facets for the first ribs and in front of this a beak-like extension, which is shorter, broader and more concave on the dorsal side than in *Felis*, and a similar posterior portion. The mesosternal segments are slender, somewhat contracted in the middle and expanded at the ends. They are relatively longer and more slender and less depressed than in the modern cats.

FORE LIMB

The *scapula*, which has lost most of the prescapular fossa, seems, nevertheless, to be very feline in character. The blade is rather short proximo-distally and broad transversely, with broad neck and no coraco-scapular notch. The spine is high and thin and ends distally in a broad antroverted acromion; there is also a metacromion, but this seems a backward extension of the acromion, above which it is but little elevated. In *Felis*, on the contrary, the metacromion is independent and is given off from the spine considerably above the distal end. The coracoid is smaller and less distinctly separate than in *Felis*.

The *humerus*, while altogether cat-like, differs in many details from that of *Felis*, especially in being relatively shorter and more slender. The head is smaller and projects less prominently behind the shaft. The external tuberosity is smaller, not rising to the level of the head, from which it is separated posteriorly by a sulcus. Anteriorly, however, it is more prolonged and incurved, making the bicipital groove narrower and deeper. The internal tuberosity is much reduced, as in *Felis*. The shaft is short and slender and differs from that of the modern genus especially in the shape of the proximal portion, which, in the latter, has a broad, flattened, triangular area for the deltoid muscle, bounded on both sides by rugose borders. In *Dinictis* there is no such area, but the anterior side of the shaft is an edge; the deltoid ridge, which extends distally to the middle of the shaft, is far less prominent than in *Drepanodon*. The supinator ridge, on the other hand, is more conspicuous and extends farther up the shaft than in *Felis*, and the internal epicondyle is much more prominent, though less massive and rugose. The epicondylar foramen is wider open, less slit-like, and the bony bridge over it is more slender. The trochlea is shaped as in *Felis*, but is much lower proximo-distally; there is an external convexity for the head of the radius and an inner, saddle-shaped surface for the ulna; there is no supra-trochlear fossa.

The forearm bones are of about the same relative length, in comparison with the humerus as they are in the Leopard, but they are much more slender. The *radius* has a discoidal head, which evidently had great freedom of rotation on the humerus, as in the cat-tribe generally; the bicipital tubercle is not so large, or massive as in *Felis*, but has a similar position. The shaft is entirely different in shape from that of the modern genus in being slender and rounded, nearly cylindrical, instead of being broad and antero-posteriorly flattened. The distal end is much expanded transversely, but less than in *Felis*, for it does not extend beneath the ulna, the facet for which presents laterally, not proximally.

The *ulna* differs from that of the Leopard in subordinate details only. The olecranon is much narrower antero-posteriorly, but rather thicker transversely and has no tendinal sulcus on the free end. The shaft, especially the proximal portion, is much more slender, though, in comparison with the radius, it cannot be called reduced. Both on the external and internal sides the shaft is channelled, the broad groove on the outer side extending for nearly the whole length of the bone. In the Leopard, there is an obscurely marked channel on the outer side, but on the internal side the shaft is convex. The distal process for articulation with the radius is less prominent than in the Recent cats and the contact is lateral. The styloid process is long and slender and has a simply convex surface for the pyramidal.

Manus

The *carpus* has, for the most part, been lost and what remains differs in no material respect from that of *Drepanodon*.

The *metacarpus* consists of five complete members, which are remarkably short in comparison with the metatarsals, and still more so as compared with those of *Felis*. The bones are of unequal lengths; in inverse order they are I, V, II, IV, III; the last two are nearly equal. The arrangement is radiating, not parallel, as it is in modern cats and dogs, retaining the more primitive arrangement seen in most creodonts and in the pentadactyl, more or less plantigrade bears, raccoons, mustelines and civets throughout their recorded history.

The *first metacarpal* is very short, much the shortest of the series and, if complete in the mounted foot, must have formed so short a pollex that it can have been only a dew-claw; the distal trochlea is very small, but has a well-formed hemispherical head.

The *second metacarpal* is nearly twice as long as the first and much stouter; the *third* is the longest and heaviest of the series; its head is overlapped by that of mc. II, which abuts laterally against the magnum and, in turn, mc. III overlaps the head of mc. IV and rests against the unciform. The *fourth metacarpal* is of nearly the same length as mc. III, but decidedly more slender. The *fifth* is very short, next after mc. I, the shortest of the series. The distal ends of all the metacarpals have hemispherical trochleae for articulation with the first row of phalanges, that are very cat-like.

The *phalanges* are short and slender and clearly show the retractility of the claws, which, however, was probably not so perfect as in existing cats. Phalanges of the second row have asymmetrical distal ends and the shank is, as it were, excavated on the ulnar side, which allows the ungual to be rotated past it into the foot. In *Dinictis* the asymmetry, though unmistakable, is less extreme than in *Felis* and it looks as though the claws could not have been completely hidden in the living foot. The *ungual* phalanges have a thin, sharp-pointed, blade-like core which is nearly straight and from the base of which arises a short bony hood that covers the core for about half its length and is thus very much smaller than in *Felis*, in which the hood is so large that, in side view, it covers and conceals the bony core of the claw. The horny part projects far out beyond the hood. The hooded claw is pre-eminently an aeluroid character and few other mammals have it; in *Dinictis* it appears in an incomplete, incipient stage.

HIND LIMB

The *pelvis* is not particularly feline in character, but is rather more like that of the Viverridae. The ilium is narrow, as in the Carnivora generally; the neck is short, deep and thick and the anterior expansion does not much exceed it in width. The gluteal surface is not simply concave, as in *Felis*, but is divided by a sharp ridge into upper and lower concavities, of which the upper is the broader. This ridge also occurs in *Pseudocynodictis* and is faintly marked in *Viverra*. The acetabular border is short, but broad, rugose and prominent and the ilio-pectineal rugosity is distinct. The ischium is long, straight, slender and of trihedral cross-section; the posterior portion is hardly at all everted, a striking difference from both *Felis* and *Canis;* the tuberosity is a mere thickening of the border and not at all everted. The spine of the ischium is represented by a slight convexity of the dorsal border, which ends abruptly behind, thus forming the lesser sacro-ischiadic notch, which has about the same position as in *Cryptoprocta*. The descending portion of the ischium is thin and plate-like and meets its fellow of the opposite side in a very long symphysis. The pubis is relatively long, broad and thin; the limits of pubis and ischium are no longer distinguishable, but the obturator foramen is, like the symphysis, very long.

The *femur* is long and slender and resembles that of many of the more primitive fissipeds. The head, which is sharply constricted from the neck and is evenly rounded, forming somewhat more than a hemisphere, projects more obliquely upward and inward than in the true cats. The great trochanter is massive and much extended from before backward, but rises only slightly above the level of the head and its proximal edge is more regularly rounded than in Recent felines; the digital fossa is small, but deep. The second trochanter is large and prominent and the intertrochanteric ridge is a short, curved, rugose line. Of particular interest is the presence of a small third trochanter, which has the same unusually proximal position as in *Daphoenus;* the third trochanter, if it can properly be called such, runs for some distance down the shaft and is continued as an external *linea aspera.* The shaft of the femur is long and slender and strongly arched forward; the anterior face is regularly curved transversely, but the posterior face is flattened, which prevents the shaft's having a cylindrical shape. The distal portion of the shaft broadens gradually to the condyles; the popliteal region is nearly smooth and has no such rugosities for muscular attachment as are seen in the larger cats. The condyles, which are of nearly equal size and are separated by a broad inter-condylar notch, project but little behind the plane of the shaft, an indication of a plantigrade gait. On the proximal faces of the condyles are small facets for the *fabellae.* The rotular trochlea is broad, shallow and symmetrical, which gives a feline appearance to this end of the bone.

The *patella* is broad, but thin, and of an almond-like shape. The articular surface for the femoral trochlea is somewhat concave proximo-distally and even more slightly convex transversely. The anterior surface and sides are but little roughened, rather they are faintly striate.

The *tibia* is considerably shorter than the femur, in the ratio of 8.4 to 10; seen from the front, the bone appears to be straight, but, viewed from the side, it has a strong forward curvature. The surfaces for the femoral condyles are flattened and almost in contact, with very low spine. The cnemial crest is not very prominent and does not descend far. On the postero-external angle of the outer condyle, there is a large, flat, oval facet for the head of the fibula. The shaft is trihedral in its upper two-thirds; the lower third is compressed and of oval cross-section. The distal end is peculiar and is but little expanded; the internal malleolus is long and very heavy and projects somewhat internally, as well as distally and the distal end is abruptly truncate. The sulcus for the tibial tendons is double, as in the felines, and the bounding ridges are continued to the distal border of the malleolus. The astragalar surface is much flattened and the inter-condylar ridge is hardly indicated, being merely the angulation at which the two surfaces meet and the dorsal tongue for the groove of the astragalus is faintly shown.

The *fibula* is comparatively well developed and very cat-like; the proximal end is considerably thickened antero-posteriorly and broadened transversely; it bears a large, oval and obliquely placed facet for the tibia. The shaft, though slender, is relatively as stout as in the cats and civets and is of irregularly trihedral shape, with sharp *crista interossea.* The distal end is broader and thicker than the proximal and, in shape, resembles that of modern felines; on its posterior border is a deep sulcus for the peroneal tendons and, on the tibial side, is a large facet for the outer side of the astragalus.

Pes

The hind foot is much longer than the fore foot, though slender and weak in comparison with the pes of a modern cat of corresponding size. The *calcaneum* has considerable resemblance to that of *Procyon;* the tuber is short and stout, with moderately expanded free end, which is grooved by a sulcus for the Achilles tendon. The external astragalar facet is very long and narrow and presents internally more than dorsally; the sustentaculum is heavy and prominent and its astragalar surface is reflected over upon the proximal edge. The facet for the cuboid is oval and concave and slopes downward toward the fibular side and on that side of the calcaneum, near the distal end, is a prominent projection for ligamentous attachment, such as occurs also in *Daphoenus* and *Pseudocynodictis* and in various Recent genera which are, for the most part, plantigrade, such as *Procyon* and *Gulo.*

The *astragalus* is remarkable for the flatness of its trochlea, the intercondylar groove being hardly more than indicated; the outer condyle is much broader and somewhat higher than the inner one and forms nearly a right angle with the very large and slightly concave surface for the fibular malleolus. At its proximal end the trochlea becomes very narrow and somewhat more deeply grooved; distally, it ends abruptly and is not continued down over the neck, as it is in *Felis*. The tibial side of the astragalus is very oblique and passes by a gentle curve into the inner condyle, quite different from the sharp edge found in most Recent fissipeds. The neck is short and not very much deflected toward the tibial side; the distal end is formed by the large convex head for the navicular, but, on the fibular side, there is a small, though distinct, facet for the cuboid, which also is found in *Daphoenus*, but rarely occurs in Recent fissipeds, the bears and certain mustelines having it. The facet for the sustentaculum is long and rather narrow, convex distally and concave proximally, where it embraces the proximal edge of the sustentaculum. The external calcaneal facet is deeply concave proximo-distally and has an unusually oblique position.

The *navicular* is uncommonly broad and very simple; its proximal end is taken up entirely by the deeply concave surface for the head of the astragalus; on the fibular side is a single narrow facet for the cuboid and the distal side has three facets, of which that for the ecto-cuneiform is much the largest and that for the ento-cuneiform is very obliquely placed.

The *cuboid* is a stout bone, with large diameters in all three dimensions; the proximal surface is somewhat convex and slopes toward the fibular side; internally from the calcaneal surface is a narrow, oblique facet for the head of the astragalus. The tibial side of the cuboid has a single small and nearly plane facet for the ecto-cuneiform. On the distal end is a very large and slightly concave surface for the head of the fourth metatarsal and, external to this, a very much narrower facet for the fifth, which has an oblique position, presenting outward as well as distally. On the fibular side is a well-marked sulcus for the peroneal tendons.

The *ecto-cuneiform* is large, very broad on the dorsal side, narrowing rapidly to the plantar face, from which is given off a stout, knob-like process. On the tibial side is a pair of small facets for the second metatarsal. The *meso-cuneiform* resembles that of *Paradoxurus* in shape, having a very oblique surface for the navicular; the tibial side of the bone is much higher proximo-distally than the fibular and, consequently, the distal face slopes steeply outward. The *ento-cuneiform* is long proximo-distally and narrow transversely, but has considerable dorso-plantar thickness; the proximal end rises sharply toward the plantar side, which makes the length greater on that side than on the dorsal. This bone extends distally beyond the level of the meso-cuneiform and articulates with the side of the second metatarsal.

As compared with those of most existing Felidae, the *metatarsals* of *Dinictis* are short and weak and have a viverrine appearance, but they are very much longer than the metacarpals, the disproportion being conspicuously greater than in *Felis*. All the members of the metatarsus are of different lengths, but the third and fourth are nearly equally long; the inverse order is I, V, II, III, IV. The hallux would seem to have been functional, though it is the shortest of the digits. The *first metatarsal* has an enlarged head, which bears a saddle-shaped surface for the ento-cuneiform, concave transversely, convex planto-dorsally and has a prominent rugosity on the tibial side. The shaft is short and strongly arched to the dorsal side and the distal end has a very small, but perfectly formed, hemispherical surface for the first phalanx. The *second metatarsal*, though much longer than the first, is short and slender. The head is wedge-shaped, narrowing to the plantar side, and, as is well-nigh universal in the fissipeds, on account of the small meso-cuneiform, it rises proximally above the adjoining metatarsals and is held between the ecto- and ento-cuneiforms. On the fibular side is a slight depression for mt. III, but mt. II is not so firmly interlocked with its neighbours as are the three on the fibular side of the foot. The shaft is moderately curved toward the dorsal side.

The *third metatarsal* is the stoutest, though not the longest, of the five, and the proximal portion is especially broad; on the fibular side of the head is a deep depression, into which fits a projection from that of mt. IV. The facet for the ecto-cuneiform is large and oblique to the long axis of the bone, sloping down toward the tibial side. The connections of the head are of the ordinary feline type; the two facets for mt. IV are separated by a deep emargination and the plantar side is very narrow. The only tarsal articulation is with the ecto-cuneiform. The *fourth metatarsal* is slightly longer and decidedly more slender than the third, but heavier than any of the others; it has a narrow, convex head for the cuboid and is firmly interlocked with the adjoining metatarsals on each side; on the tibial side is the rounded prominence, already mentioned, which is received into the pit on mt. III, and on the fibular side is a similar pit which receives a projection from mt. V. The shaft of mt. IV is curved, so that the distal end is deflected outward and is slightly twisted upon itself. The *fifth metatarsal* is of nearly the same length as mt. II and a little more slender; the surface for the cuboid is very narrow and convex planto-dorsally and above it rises the large, thickened and rugose projection from the fibular side, which is much the same as in *Felis;* the shaft is curved inward and dorsally, like mt. IV.

The distal ends of all the metatarsals have rounded, hemispherical heads for the phalanges and these are sharply constricted from the shafts, and on the plantar side are thin and prominent keels. Proximal to these articular hemispheres, the distal ends of the shafts are broadened and have rugose processes for the attachment of the lateral ligaments and above each hemisphere is a conspicuous fossa.

The *phalanges* are feline in shape, but relatively shorter and more slender than those of modern cats, and resemble those of *Cryptoprocta*, though somewhat stouter than in that genus. Those of the proximal row are comparatively broad and strongly arched toward the dorsal side; those of the second row are much shorter, more depressed and flattened and near the distal end, on the fibular side, is a depression to allow the retraction of the claws, but this is much less marked than in *Felis*, though more so than in *Cryptoprocta* and most of the viverrines. *Prionodon (fide* Mivart) has claws almost as completely retractile as those of the cats, but this can hardly have been true of *Dinictis*, though, unquestionably, there must have been a certain degree of retractility. The *unguals* are very small, short, thin and sharp-pointed and nearly straight; indeed, they are much like the unguals of *Paradoxurus*, but the articular surface is of a different shape and the subungual process is larger, though very much smaller than in modern cats. The hood, which in the manus, is complete, though very short, leaving most of the claw-core exposed, is in the pes very little developed and in an incipient stage. Such a difference between fore- and hind-feet is surprising, though a difference in the number of digits is not very uncommon.

  · *Species.* The species of *Dinictis* are in a state of confusion, for the clearing up of which a much larger suite of skulls and skeletons than has yet been collected will be required.

Dinictis felina Leidy

˙(Pls. XV, XXII, Figs. 3, 4)

Dinictis felina Leidy, Proc. Acad: Nat. Sci. Phila., 1854, p. 127.
Deinictis felina Leidy, *ibid.*, 1856, p. 91.

The typical species of the genus is distinguished by its medium size, by the obliquity of the brain-case and by the relatively large size of the glenoid and mastoid pedicles. That is to say, these processes are large for a species of this genus, but they have no such development as in *Drepanodon* and they are incomparably smaller than in *Eusmilus*, in which they attain an excessive size.˙ There are two well-defined species of the genus in the White River, *D. felina* and *D. squalidens*, of which the latter is typically smaller than the former, but is connected with it by so many intergradations, that a separation founded on size alone would be of doubtful propriety. Much more significant are the differences in skull structure, *D. squalidens* inclining to the Nimravinae in the basi-cranial characters. So marked is the difference, that even a generic separation may be found necessary. The lumbar and caudal vertebrae are much heavier, with longer and stouter neural spines and transverse processes than in *D. squalidens* and the tail is considerably longer.

In the subjoined table only skull dimensions are given, because lack of time made it impossible for us to measure the fine skeletons, attributed to this species, in Pittsburgh, Chicago and Rapid City; dimensions of the Cambridge skeleton are given under *D. squalidens*.

In this table the measurements of teeth are taken from a skull in the Princeton Museum (No. 10,012) and those of the skull are from one of Leidy's specimens in the Academy of Natural Sciences, Philadelphia.

MEASUREMENTS

Upper incisor series, width	23.0 mm.	Lower canine, transv. diam.	10.5 mm.
I2, ant.-post. diameter	4.5	Lower cheek-teeth series, length	52.0
I2, transverse diameter	3.0	P$\overline{2}$, ant.-post. diameter	6.0
I3, ant.-post. diameter	6.5	P$\overline{3}$, ant.-post. diameter	12.0
I3, transverse diameter	5.0	P$\overline{4}$, ant.-post. diameter	14.5
Upper canine, ant.-post. diam.	15.0	M$\overline{1}$, ant.-post. diameter	18.5
Upper canine, transverse diameter	11.0	M$\overline{2}$, ant.-post. diameter	6.0
Upper cheek-teeth series, length	47.0	Skull, length occ. cond. to prmx.	154.0
P2, ant.-post. diameter	5.0	Cranium, length cond. to orb.	108.0
P3, ant.-post. diameter	16.0	Cranium, length orb. to constrict.	50.0
P4, ant.-post. diameter	24.0	Face, length orb. to prmx.	46.0
M1, ant.-post. diameter	4.0	Hard palate, length	69.0
M1, transverse diameter	7.0	Hard palate, width post. bord.	72.0
Lower incisor series, width	19.0	Hard palate, width at canines	26.0
I$\overline{3}$, ant.-post. diameter	4.5	Distance *for. mag.* to post-glen.	33.0
I$\overline{3}$, transverse diameter	3.0	Distance *for. mag.* to mast.	19.6
Lower canine, ant.-post. diameter	15.0	Mandible, length fr. cond.	119.0

Dinictis squalidens Cope

(Pls. XVII, XVIII)

Daptophilus squalidens Cope, Palaeont. Bull., no. 16, p. 2 (1873).
Dinictis squalidens Cope, Proc. Acad. Nat. Sci. Phila., 1879, p. 76.

On the whole, this is the smallest species of the genus, but there is great variability in the matter of size.   Much more significant are the horizontal brain-case and the small size, almost absence, of the glenoid and mastoid pedicles.

In the table the measurements are taken from the Harvard skeleton and, for comparison, those of a full grown African Leopard are given.   The second column, to the right, under each is of the proportional measures, the length of skull being taken as 100.

MEASUREMENTS

	F. pardus		*M. C. Z.*	
Skull, length occ. cond. to prmx.	192.0 mm.,	100.0	151.0 mm.,	100.0
Cranium, length cond. to orbit	136.0	70.8	107.0	70.8
Face, length, orb. to prmx.	56.0	28.1	? 44.0	? 28.0
Orb. to post-orb. constr., distance	48.0	25.1	48.0	31.1
Mandible, length from condyle	135.0	73.1	112.0	73.6
Neck, length	200.0	100.4	121.0	80.1
Axis, ant.-post. breadth of spine	60.0	31.2	44.0	29.1
Thorax, length			235.0	15.1
Loins, length	239.0	124.5	218.0	144.4
1st dorsal, length	19.0		18.0	
13th dorsal, length	24.0		26.0	
1st lumbar, length of centrum	26.0		26.0	
3rd lumbar, length of centrum	34.0		30.0	
5th lumbar, length of centrum	35.0		34.0	
6th lumbar, length of centrum	37.0		33.0	
1st caudal, length of centrum	29.0		15.0	
1st caudal, width over trans. proc.	51.0		42.0	
Middle caudal, length	42.0		27.0	
Middle caudal, dors.-vent. thickness	14.0		10.0	
Scapula, prox.-dist. length	155.0	80.7	109.0	72.1
Humerus, length from head	211.0	109.8	144.0	95.3
Humerus, width of prox. end	42.0	21.8	29.0	19.2
Humerus, width of dist. end	49.0	27.1	31.0	20.5
Radius, length	191.0	100.0	128.0	84.2
Radius, width of prox. end	21.0		20.0	
Radius, width of dist. end	34.0		25.0	
Ulna, length	237.0	123.4	158.0	104.6
Ulna, length of olecranon	33.0		29.0	
Ulna, width at humeral facet	26.0		21.0	
Ulna, width of distal end	9.0		5.0	
Metacarpal II, length	66.0	34.3	20.0	13.2
Metacarpal II, width prox. end	12.0		6.0	
Metacarpal III, length	74.0	38.5	28.0	18.5
Metacarpal III, width prox. end	14.0		14.0	
Metacarpal III, width dist. end	14.0		10.0	
Metacarpal IV, length	68.0	35.9	30.0	19.2
Metacarpal IV, width prox. end	16.0		9.0	
Metacarpal IV, width dist. end	12.5		8.0	
Metacarpal V, length	59.0	30.7	24.0	13.9

	F. pardus		M. C. Z.	
Metacarpal V, width prox. end........	16.0		9.0	
Metacarpal V, width dist. end........	12.5		8.0	
Manus, digit IV, 1st phal., length.....	35.0	18.2	18.0	11.9
Manus, digit IV, 2nd phal., length.....	29.0	15.6	16.0	10.6
Manus, digit IV, 3rd phal. length.....			12.0	
Ischium, length.....................	76.0	39.6	65.0	43.0
Pubic symphysis, length.............	66.0	34.4	57.0	37.4
Obturator foramen, length............	44.0	23.9	44.0	29.1
Femur, length from head.............	233.0	121.4	174.0	115.3
Femur, width prox. end...............	47.0		33.0	
Femur, width distal end..............	42.0		28.0	
Tibia, length on inner side...........	223.0	116.2	151.0	100.0
Tibia, width prox. end...............	47.0		32.0	
Tibia, width dist. end................	35.0		19.0	
Fibula, length......................	208.0	108.3	140.0	92.7
Fibula, thickness prox. end...........	17.0		12.0	
Fibula, thickness dist. end...........	19.0		15.0	
Calcaneum, length..................	62.0	32.3	42.0	27.8
Astragalus, length..................	34.0	22.5	25.0	16.2
Metatarsal I, length.................			22.0	
Metatarsal I, width prox. end.........			6.0	
Metatarsal II, length...............	82.0	42.7	45.0	29.1
Metatarsal II, width prox. end........	11.0		6.0	
Metatarsal III, length..............	91.0	47.4	45.0	29.8
Metatarsal III, width prox. end.......	15.0		11.0	
Metatarsal IV, length...............	92.0		46.0	
Metatarsal IV, width prox. end.......	12.0		8.0	
Metatarsal V, length................	80.0	41.1	37.0	24.5
Metatarsal V, width prox. end........	35.0		18.0	
Pes, digit IV, 1st phal. length........	35.0		18.0	
Pes, digit IV, 2nd phal. length........	29.0		16.0	
Pes, digit IV, 3rd phal. length.......			12.0	

These figures make clear the relative size and proportions of this skeleton; the head is small, the neck short, the trunk and especially the loins, long, but light, and the tail notably shorter and thinner than in the Leopard and in the White River sabre-tooth *Drepanodon*. As compared with those of the Leopard, the limbs are short and slender, the feet small and weak.

Horizon: lower Brulé.

Several additional species of this genus have been named, but their status is very doubtful and must await the discovery of far more complete and extensive material than has yet been collected.

Dinictis bombifrons Adams

Dinictis bombifrons Adams, Amer. Natur., XXX, p. 577, 1895.

Dinictis fortis Adams

Dinictis fortis Adams, *ibid.*

Dinictis paucidens Riggs

Dinictis paucidens Riggs, Kansas Univ. Quarterly, IV, pp. 237–41, 1896.

Dinictis cismontanus (Thorpe)

Pogonodon cismontanus Thorpe, Amer. Journ. Sci. (4), L, p. 222, 1920.

Dinictis major Lucas

Dinictis major Lucas, Am. Journ. Sci. (4), VI, p. 399, 1898.

The original description of this supposed species is: "The distinctive characters are the size of the animal, the feeble development of the mandibular flange for the protection of the upper canine, the robust character of the feet and the presence of an ungual shield (or hood)." Unless there was an accidental association of the foot-bones of *Drepanodon* with the mandible of *Dinictis*, this species must be accepted as distinct.

Drepanodon Leidy

(Pls. XV, XVI, XVII, XIX)

Machairodus Leidy and Owen (*nec* Kaup) Proc. Acad. Nat. Sci. Phila., 1851, p. 329.
Drepanodon Leidy (*nec* Nesti), *ibid.*, 1857, p. 176.
Hoplophoneus Cope, Bull. U. S. Geolog. and Geogr. Surv. Terrs., No. 1, p. 23, 1874.
Dinotomius Williston, Kansas Univ. Quart., 1895, p. 170.
Drepanodon Palmer, T. S., Index Gener. Mamm., p. 244 (1904).

It is with much regret that we feel compelled to abandon Cope's term of *Hoplophoneus*, which has been in familiar use since 1874, but as Palmer has shown, the law of priority requires that Leidy's term should be revived. In 1851 Leidy and Owen described *Machairodus primaevus*, a species which Leidy in 1857 transferred to *Drepanodon* Nesti, 1826, in the belief that this was a recognized European genus which antedated Kaup's *Machairodus* of 1833. Palmer, however, writes: "Nesti, usually given as the authority for *Drepanodon*, merely used the name in 1826 specifically. Leidy, among others, refers the name to him and gives as synonyms of *Drepanodon:* '*Megantereon* Croiz. 1828; *Agnototherium*, *Machairodus* Kaup, 1833; *Steneodon* Croiz. 1833; *Smilodon* Lund, 1841, etc.'" (*loc. cit.*). Leidy was thus the first to use the term generically and the type species is the American *D. primaevus*.

This is much the commonest of the White River sabre-tooth genera and is more advanced and specialized than *Dinictis*. The genus, which seems to have been an immigrant from the Old World, extended throughout the White River and John Day stages but has not been found in the Miocene and cannot be regarded as ancestral to the Pliocene or Pleistocene smilodonts. Many species have been described and, though most of these will probably prove to be synonyms, there are several well-distinguished types, which ranged in size from a Lynx to a Leopard.

DENTITION

The dental formula is: $i\frac{3}{3}$, $c\frac{1}{1}$, $p\frac{3}{2}$, $m\frac{1}{1}$, though in old skulls the anterior premolar (p2) has been shed and the alveolus obliterated.

Upper Teeth. The *incisors* increase in length and diameter of crown from the median (i1) to the lateral (i3), but there is no such great preponderance of i3 as in *Felis*, and the teeth are simple, sharp-pointed, recurved hooks, which are relatively longer and more acutely pointed than in the modern genus and the posterior cingulum is less developed.

The *canine* is very long, recurved and scimitar-like, broad antero-posteriorly, thin and blade-like, with finely serrate cutting edges. In *Drepanodon* the sabre is much longer and more curved than in *Dinictis* and still more than in *Nimravus*, but smaller than in *Eusmilus*.

The manner in which the great scimitar-like tusks of this and the other sabre-tooth genera can have been employed, is difficult to understand, for the mode of biting practised by the true felines was obviously unfeasible. In order to clear the points of the sabres, the mouth must have been opened much more widely than is possible to any of the existing cats. To drop the mandible so far, a special arrangement of the temporal, masseter and sterno-mastoid muscles was necessary, and this, in turn, required particular modifications of the bones to which these muscles were attached. The manner in which the glenoid cavity of the squamosal was carried down to a low level, the position of the mandibular condyle and the size and form of the coronoid process, are all indications that these partic-ular modifications of the jaw-muscles did actually exist in the living animal.

As suggested by Dr. J. C. Merriam and the late Professor W. D. Matthew, the ma-chairodonts must have dropped the lower jaw so far as to expose the points of the upper canines, which then delivered a stabbing blow, as a venomous snake strikes with his fangs, and terrible wounds must have been inflicted. The stabbing strokes were evidently made by a violent downward movement of the whole head, actuated chiefly by the sterno-mastoid muscles and this would explain the extraordinary development of the mastoid processes in the sabre-tooths, so conspicuously different from anything seen in the true cats. This may seem like an awkward, ineffective method of killing prey, yet it evidently was highly efficient, for it spread all over the world, except Australia and, perhaps, Africa, and persisted for millions of years, from the Oligocene to the end of the Pleistocene. Why it should have eventually disappeared so completely, one can only conjecture.

The stabbing hypothesis is supported by the skull of *Nimravus* described on p. 148, which shows a dreadful wound certainly inflicted on the living animal and, most probably by the sabre of *Eusmilus* (*q.v.*) (Pl. XXI, Fig. 15).

The anterior *premolar* (p2) is very small and is frequently shed; it has a simple, conical crown and is supported on two roots. In linear dimensions, p3 is nearly three times as large as p2 and is like a miniature sectorial in shape. The sectorial, p4, is rather small proportionally, but of feline character; nearly half of the longitudinal diameter is formed by the anterior, compressed-conical cusp and the posterior, trenchant blade is of nearly the same length. The small, antero-external, accessory cusp, of which *Dinictis* has no trace and which is so characteristic of almost all cats, is present, but in an incipient and variable stage of development. The internal cusp, or deuterocone, is even more reduced than in *Felis* and is not demarcated from the external cusp, upon which it is merely a but-tress, though carried on a separate root. The only upper molar (m1) is a very small trans-verse tooth, inserted by two roots, internal and external, and when the skull is viewed from the side, it is visible externally. The crown consists of an outer cone and a transverse ridge and is much narrower, antero-posteriorly on the lingual side. Small as this tooth is, it is actually as large as in the Lion and therefore very much larger proportionally. In the latter m1 is overlapped and concealed, partly by the blade of the sectorial, partly by the alveolar process.

Lower Teeth. The *incisors* are closely crowded together and are smaller than those of the upper jaw, but of similar shape, simple, acute and slightly recurved hooks. The two median incisors (i$\overline{1}$) are very small and closely appressed; i$\overline{2}$ and i$\overline{3}$ are progressively larger. The roots of all the incisors are in the same straight line, as in the cats generally, that of i$\overline{2}$ not being displaced backward, as it is in almost all other Fissipedia. The *canine* is small; as was pointed out under *Dinictis*, upper and lower canines stand in inverse ratio to each other and in *Drepanodon*, the lower one, though far smaller than the upper, greatly exceeds i$\overline{3}$ in size, while resembling it in form. The lower canine and third upper incisor together form an efficient secondary grasping apparatus, which might be used when the mouth was not open so far as to free the points of the great upper tusks. These could hardly have been used for seizing prey and were probably exclusively stabbing weapons.

The two *premolars* are similar in form, but differ in size; p$\overline{3}$ is very narrow, compressed and trenchant; the principal cusp is sharp-pointed and has a cutting blade behind it and there is also a minute, anterior basal cuspule. P$\overline{4}$ is very much larger and plainly triconodont; the high, sharp-pointed main cusp, is placed between anterior and posterior trenchant cusps, which are lower and of equal size; the three are in the same fore-and-aft line. The first *molar* (m$\overline{1}$) is a very cat-like sectorial and consists of two trenchant blades, which are in close juxtaposition, thus differing from those of *Nimravus*. The inner cusp of the original triangle has been lost without leaving a trace and, of the heel, only a minute vestige remains, which is sometimes duplicated by an even smaller cuspule.

All the cheek-teeth of this genus above and below are far thinner and less massive, proportionally, than those of the Lion and can have been little, or not at all used for the crushing of bones, a process which the sabres would have prevented.

SKULL

The skull has an unmistakable resemblance to that of the true cats, but in many ways is much more primitive and has features which it shares with *Daphoenus* and other early genera of the Canidae. The cranium is long, the face very short, the facial shortening being more decided in some species than in others. The upper contour of the skull is much as in *Felis*, though that of the face is longer and straighter. The sagittal crest is longer and higher posteriorly and there is no such lyrate sagittal area as appears in the Lion; the temporal ridges are obscurely marked, almost obsolete indeed. The occipital crest is far more prominent, a feature, which like the height of the sagittal crest is produced by the less capacious brain-case. The cerebral fossa is less prolonged posteriorly and the postorbital constriction is decidedly deeper and placed farther behind the orbits, a primitive character.

The occipital surface is relatively higher and narrower than in the Lion and there is no roughening of the supra-occipital for attachment of the *ligamentum nuchae*. The exoccipitals are less deeply concave and the paroccipital processes are conspicuously smaller and less prominent. Indeed, this region of the skull presents one of the most striking and constant differences between the sabre-tooth genera and the true cats. In the machairodonts the small paroccipital processes are not in contact with the auditory bullae, but are directed away from them, a radical difference from the felines, in which the bulla is received into the broad, concave paroccipital "like an acorn in its cup." Another very striking difference between the two subfamilies is in the presence in the skull of the sabre-tooths of a very conspicuous pedicle formed by the conjoined mastoid and post-tympanic processes. This pedicle is parallel to an anterior one formed by the downward projection of the zygomatic root, carrying the glenoid cavity; between the two pedicles is the entrance to the ear.

The auditory bullae are relatively small and very fragile; in a large series of skulls of this and other Oligocene machairodonts, preserved in the various museums, we have not seen any in which a bulla was preserved intact. In one skull of *Drepanodon* and one of *Nimravus*, what appear to be fragments of the bulla are attached to the internal cast of the chamber. In another example of *Drepanodon* in the American Museum, the remarkably perfect skull of a young adult, there is, on each side, a tympanic ring surrounding the auditory meatus and an internal cast of the bulla. These various appearances are difficult to harmonize, or to interpret; apparently, the tympanic was ring-like and attached to it was an entotympanic chamber, which was usually cartilaginous and, when ossified, was of egg-shell fragility. In the type-skull of *D. oharrai* the bony part of the bullae shows that they were each divided into two chambers, with an arrangement more as in Viverridae than in Felidae.

The *parietals*, which form nearly all of the cranial roof, support a sagittal crest which is longer than in some modern cats (*e.g., Felis pardalis*) and higher than in any of them, because of the smaller brain-case and higher occiput. The squamosal is large and makes up most of the side-wall of the cranium. The peculiar character of the united post-tympanic and mastoid processes was mentioned above and it may be added that this pedicle is nearer to the postglenoid process than in the true felines. Between the postglenoid and post-tympanic processes the squamosal forms a projecting shelf that is wider than in modern cats. The zygomatic process curves out as boldly from the side of the skull as it does in the Recent felines, but the portion on which is placed the glenoid cavity, projects far ventrally, much more below the basicranial axis than in *Felis*, but not nearly so far as in *Eusmilus* (*q.v.*). The depth to which this projection descends is plainly correlated with the length of the canine tusks and serves to lower the level of the mandibular condyle, thus facilitating the extraordinarily wide opening of the mouth so often referred to. The zygomatic arch, as a whole, resembles generally that of the smaller modern cats, but the postorbital process of the jugal is far less prominent than in *Felis* and is little more than an angulation, leaving the orbit widely open behind.

The *frontals* are smaller than in *Felis* and are almost entirely excluded from the roof of the cerebral fossa by the elongate parietals and the postorbital constriction, which in the modern genus affects only the frontals and is well in advance of the coronal suture, is in *Drepanodon* almost coincidental with the suture. The forehead is more decidedly convex than in Recent cats, which may be due to larger frontal sinuses, or more developed ethmo-turbinals; the postorbital processes, like those of the jugals, are short. The nasal processes of the frontals are likewise short. The *nasals* differ in several details from those of existing cats; they are narrower and broaden anteriorly less; they are also flatter, not curving downward at the sides, and the anterior end of each nasal is more deeply emarginated and projects in a median point.

The *premaxillaries* differ from those of *Felis* only in having a relatively larger alveolar portion, which projects farther in front of the canines. The ascending ramus is proportionately longer and not so slender, though, owing to the extreme shortening of the nasal processes of the frontals, the premaxillae are widely separated from the latter. The palatine processes are very like those of *Felis*, though the spines are more slender and the vascular grooves leading forward from the incisive foramina are shorter and shallower. The *maxillaries* vary somewhat in the different species of *Drepanodon* in accordance with the length of the preorbital part of the face. In general, they are closely similar to the same bones in *Felis*, but the protuberances made by the sheaths of the canine tusks are less prominent. The infraorbital foramen has nearly the same position, but is more widely open. The palatine processes have almost the same shape as in *Felis*, as have also the palatine bones.

The *mandible* is highly characteristic of the subfamily, the various genera differing in the degree to which the peculiar features are developed and emphasized. In *Drepanodon* these characteristics are very striking and conspicuous, far more so than in *Dinictis*, but far less so than in *Eusmilus*. In all of the genera the anterior and lateral surfaces are sharply set off from each other by prominent angulations which, descending, form the anterior borders of the protective flanges, given off from the ventral border at the forward ends of the horizontal rami. The size of these flanges and the degree of their projection below the horizontal rami are, except in the Pliocene and Pleistocene genera, in correlation with the length of the sabres. That the flanges did actually protect the tusks is indicated by the fact that very few sabres broken in the lifetime of the animal and with the fractured end smoothed and rounded by subsequent wear, have been reported, except in the case of the type of *D. oharrai*. This is in remarkable contrast to the great Pleistocene "Sabre-tooth Tiger," *Smilodon californicus*, of which such an astonishing number of skulls have been found in the tar-pits of Rancho La Brea. Many of these skulls have sabres broken off during life, as is proved by the manner in which the broken end has been abraded. *Smilodon* has very large and elongate tusks, but only small protective flanges. It certainly seems probable that the very large flanges of *Drepanodon* saved the canines from fracture. The outer side of the flange is concave for the reception of the upper canines, when the jaws are occluded.

It should be noted that the same protective structure has been developed in two other mammalian orders, which are as widely removed from each other and from the fissiped Carnivora as orders well can be. One of these is the extraordinary predaceous marsupial, *Thylacosmilus*, discovered by Mr. E. S. Riggs, of the Field Museum, in the Pliocene of Argentina. In this remarkable animal the sabres and protective jaw-flanges are even more extravagantly developed than in the Oligocene machairodont *Eusmilus*. The second group, the Dinocerata, is an extinct suborder of hoofed animals, confined to the Paleocene and Eocene of North America and eastern Asia. In the males of some of these genera the upper canine is a long and broad, straight, bayonet-like tusk, pointing directly downward, in other genera the tusks are curved sabres, and just such a pair of flanges are on the lower jaw as appear in the fissiped *Drepanodon* and the marsupial *Thylacosmilus*. That this similar structure originated independently in three separate groups of mammals can hardly be doubted.

Except for the bony flange and the shape of the chin, the horizontal *ramus mandibuli* in *Drepanodon* has a shape like that displayed by *Felis*, but the post-dentary portion is relatively shorter and the angular hook more slender. The masseteric fossa is deeply impressed, but covers a smaller area than in the modern genera of the family, and the coronoid process is far smaller and weaker than in *Felis* and this, in turn, would indicate a reduction of the temporal muscles. The great enlargement of the mastoid process, on the other hand, suggests a corresponding increase in the strength of the sterno-mastoid muscle. These facts are distinctly favourable to the Merriam-Matthew hypothesis of a stabbing blow, delivered by the whole head.

Cranial Foramina. So far as they are determinable, these foramina are not particularly aeluroid, but are rather of the primitive, undifferentiated character found in the earlier members of most of the fissiped families, notably that of the dogs. The number and position of the various foramina are nearly identical with those of *Dinictis;* the condylar foramen is not enclosed in the *foramen lacerum posterius,* an alisphenoid canal is present and the *foramen rotundum* and *foramen ovale* have a more posterior position than in *Felis.*

VERTEBRAE, RIBS AND STERNUM

The vertebral formula is: C 7, D 13, L 7, S 3, Cd 21, the same as in *Felis,* but the relative lengths of the vertebrae in the different regions are not the same as in any of the modern species of the family. The neck is relatively short and the vertebrae weak. The *atlas* differs from that of existing cats in having the transverse processes much narrower antero-posteriorly and the atlanteo-diapophysial notch converted into a foramen. The *axis* has a similarly large neural spine as that of *Dinictis* and *Daphoenus,* which is prolonged much farther behind the neural arch than in *Felis* and has quite a different shape. The other *cervical vertebrae* show many differences of detail from those of the large modern cats, *e.g.,* the Lion; the transverse processes of the axis and the third vertebra are more slender and the neural spines of all the postaxial cervicals are higher. On the neural arch, in the third, fourth and fifth vertebrae, there is on each side of the neural spine and internal to the postzygapophysis, a short, very distinct and backwardly directed spine-like process. Of these only vestiges are to be seen on the corresponding vertebrae of the Lion. In the latter, there is on each side of the neural arch of cervicals 3 to 6, inclusive, a large, deep, oval fossa, but of this conspicuous feature, not a trace is visible in the fossil.

The *dorsal vertebrae* do not display any very characteristic features; the neural spines are relatively thinner and weaker than in the large Recent species of the family and are of more regular shape. Instead of having one anticlinal vertebra, as is the general rule, *Drepanodon* has three, the eleventh, twelfth and thirteenth dorsals having erect spines. Anapophyses are present, though very small, on the same three vertebrae and metapophyses on the twelfth and thirteenth.

The *lumbar vertebrae* are very feline in character and, though the centra are not so slender as in *Dinictis,* they are not nearly so heavy as in the Recent species of corresponding or larger size. Anapophyses and metapophyses are shorter and not so heavy and the neural spines are lighter. The transverse processes are longer and stouter than in *Dinictis,* but not nearly so elongate as in such a Recent cat as the Leopard, for example.

The *sacrum* consists, as is usual in the cats, of three vertebrae, of which only the first supports the pelvis; the three neural spines are separate and well-developed, a difference from the Lion, in which the spine of the third sacral vertebra has been suppressed. The centrum of the third vertebra, as conditioned by the development of the tail, is still large.

The *caudal vertebrae,* 21 in number, are very Leopard-like, especially in the larger and heavier species of *Drepanodon.* The first three caudals have well-developed neural spines and the first six have single transverse processes. In the middle region of the tail the vertebrae are decidedly stouter than the corresponding ones of *Dinictis*, in which the tail must have been shorter and more slender. In *Drepanodon* the vertebrae of the middle part of the tail are very much like those of the Leopard, both in size and in the development of the various processes, and from the middle region to the end of the tail, the vertebrae follow a similar gradual reduction.

The *ribs* are short, relatively stout, rounded and rod-like, though distinctly heavier than the slender ribs of *Dinictis;* they are less flattened and grooved on the anterior face than in the Lion. The first rib is like that of the latter species, shorter, broader, more compressed laterally and plate-like than the succeeding ones.

The *sternum*, so far as it is preserved, is made up of segments which are rather heavier and stouter than in *Dinictis*. On the other hand, they are of more uniform diameter, less constricted in the middle and expanded at the ends than in the modern cats.

FORE LIMB

The *scapula* is very feline in character; as in the Lion, the pre-scapular fossa is wider than the post-scapular; the spine is high and somewhat recurved and the acromion is very prominent, descending to the level of the glenoid cavity. There is no metacromion and this is a conspicuous difference both from *Dinictis* and *Felis*. Another difference from the latter is the reduced coracoid, which is a mere rugosity.

The *humerus* differs from that of the large Recent cats in many particulars. The external tuberosity is relatively smaller than in the Lion and inclined more toward the radial side; the internal tuberosity is likewise smaller, making the bicipital groove narrower and shallower and presenting more inward, less directly forward. The shaft is less rounded, more compressed laterally and the deltoid ridge is much more prominent and extends farther down the shaft, ending distally in a distinct angulation. The distal part of the shaft is wider proportionately than in *F. leo*, the supinator ridge more prominent and the internal epicondyle much larger; the foramen is present, as in the cats generally. The trochlea is nearly the same as in the modern animal, but the internal border makes a sharper flange.

Compared with the humerus of *Drepanodon*, that of *Dinictis* is in all ways more primitive and less specialized and, in particular, less muscular. The most conspicuous difference is in the development of the deltoid ridge, so short and low in the latter, so long and prominent in the former. Another marked difference is in the distal trochlea, which is higher proximo-distally in *Drepanodon*.

The *radius* is very like that of existing cats, but its proportions differ widely in the various species of *Drepanodon:* in *D. primaevus*, the bone is short and slender. The smaller species, *D. oreodontis*, has a relatively longer forearm and the largest species, *D. occidentalis*, has a very heavy radius, which actually, as well as proportionally, is stouter than in a full-grown Leopard, and the head is so broad that there can have been little power of rotation.

The *ulna* is more distinctly different from that of a large modern cat than is the radius, though the differences are unimportant. In *Drepanodon* the posterior border of the olecranon and shaft are more convex and the olecranon itself is narrower antero-posteriorly and has no tendinal sulcus. The distal end of the ulna is entirely feline in shape.

MANUS

The fore foot is relatively small and weak, as compared with one of the larger species of *Felis*, and the *carpus* differs from that of the modern genus only in insignificant details. The *scapho-lunar*, which is made up of the ankylosed scaphoid, lunar and central, is much the largest of the carpal elements; it rests upon the trapezium, trapezoid and magnum, but has only a lateral contact with the unciform. In *Felis* this contact is more oblique, so that the scapho-lunar rests, to a certain extent, upon the unciform. As in the cats generally, the *trapezoid* is much larger than the small *magnum*. The *unciform* differs considerably in shape from that of the large modern felines, being more nearly pyramidal and having a triangular dorsal face, with the apex upward. In *F. leo*, for example, the unciform is an irregular pentagon, the sides of which are the facets for the pyramidal, scapho-lunar and magnum, the metacarpals and the ulnar border, meeting at distinct angles. The magnum also is of a different shape in the modern species, being of more uniform proximo-distal diameter.

The *metacarpals*, five in number, are much shorter than the metatarsals and are proportionally shorter and stouter than in the Lion. In *Dinictis* the metacarpals are strikingly weaker and more slender than they are in *Drepanodon*, in which the shortness is a more conspicuous feature than the slenderness; in each digit the combined length of the phalanges exceeds that of the metacarpal. The *pollex* is less reduced and relatively longer than in *Felis* and the first metacarpal has a different mode of articulation with the trapezium; in the modern genus this contact is lateral, but in the fossil mc. I meets the distal face of the carpal. The digit must have been functional.

The next two metacarpals, mc. II and III increase regularly in length, mc. III being the longest and stoutest of the series. The proximal end of mc. II does not extend over the head of mc. III and has no contact with the magnum, while in *Felis* this contact is made; its facet for mc. III is but slightly oblique. Mc. III has a very small articulation with the trapezoid, from which it is not excluded, as it is in *Felis*, by the contact of mc. II with the magnum; mc. III further differs from that of the modern genus in having a more extensive articulation with the unciform. Mc. IV is somewhat shorter than III, but their distal ends are at nearly the same level, because mc. IV does not extend quite so far proximally into the carpus. The shaft of mc. IV is more slender than III, or even than II. Mc. V is much shorter in proportion to mc. IV than it is in *Felis*.

All of the metacarpals have the characteristic hemispherical distal trochleae of the cat family, but the processes on each side of the trochlea for the attachment of ligaments, are very much less prominent than in the large modern cats; this is a conspicuous difference.

The *phalanges* are typically feline, but with several deviations from those of the Recent genera of the family. The phalanges of the proximal row are relatively more slender, but, otherwise, similar to those of *Felis*. Those of the second row are asymmetrical, to permit the retraction of the claws, but very much less so than in the modern cats, a clear indication that the claws were less perfectly retractile than in the latter, though more so than in *Dinictis*. The *unguals* are hooded, the hoods being much longer and more complete than in the genus last-named, concealing the claw-core in side view, but decidedly shorter than in modern cats, and the subungual processes are smaller.

Altogether, the manus of *Drepanodon*, while of unmistakably feline type, is yet distinctly less advanced and specialized than it is in the existing members of the family.

Hind Limb

The *femur*, in relation to the length of the skull, is somewhat longer in *Drepanodon* than in the Lion and, in the various species of the former, its proportions differ greatly. For example, in *D. primaevus* it is slender and in *D. occidentalis* very stout. In general, the great trochanter is smaller and less extended antero-posteriorly than in the Lion, but rises higher proximally and makes a more distinct sigmoid notch; the second trochanter is more prominent and farther removed from the head. A low, but distinct rugosity placed high up on the outer side of the shaft and connected with the great trochanter by a ridge, recalls the entirely similar structure in *Dinictis* and *Daphoenus* and may, as in them, be a vestige of the third trochanter, but, if so, it occupies an exceptionally elevated position. Nothing of the sort is visible on the femur of existing cats or dogs. In uncrushed specimens the shaft of the femur is cylindrical and the distal end resembles that of the larger Recent cats, though with a number of differences. In the Lion, for instance, the lateral epicondyles and the rotular trochlea are much more prominent than in the fossil.

The *tibia*, like the femur, differs decidedly in its proportions in the various species of *Drepanodon*, but the structural features are very uniform throughout the genus. There is a general similarity to the tibia of the large modern cats, though with some not unimportant differences. The bifid spine is very obscure, with its two parts in contact, and the condyles differ from those of the modern species in their inequality, the internal one being more concave and the external one convex. The cnemial crest is much less prominent and massively rugose than in *F. leo* and the distal end of the bone is of more primitive character. The internal malleolus is well-developed, but the groove for the inner condyle of the astragalus is far shallower, while the dorsal tongue and the intercondylar ridge, which fits into the groove of the astragalus, are very much less developed, though representing an advance over *Dinictis* in these respects.

The *fibula* is slender, though relatively rather stouter than in the Lion; it is also straighter, the interosseous space being wider and more uniform. The ends, especially the proximal one, are smaller than in the living species; the distal end is broadened, to form the external malleolus, but it is narrower than in *Felis* and has no sulcus for the peroneal tendons.

Pes

The hind foot is cat-like, yet with many differences of detail from that of existing species. The *astragalus* has a narrower and flatter trochlea, with so shallow a groove, as to suggest a plantigrade gait, though, perhaps, semi-plantigrade would be a better term. The astragalus of *Dinictis* is even flatter and less grooved than in *Drepanodon* and we regard it as probable that the former was truly plantigrade, though in all the museums which are so fortunate as to possess skeletons of *Dinictis*, these are mounted with a semi-digitigrade pose, which is probably not far from the truth as regards *Drepanodon*. The neck of the astragalus is longer in the latter than in *Felis* and the distal head somewhat shallower.

The *calcaneum* is short and stout; the sustentaculum is smaller than in *Felis* and the tuber calcis somewhat shorter and more compressed; a tendinal sulcus may, or may not be present on the free end. The conspicuous projection from the fibular side of the calcaneum, near the distal end, which occurs in *Dinictis*, *Daphoenus* and the plantigrade fissipeds generally, is not displayed in *Drepanodon*.

The *navicular* has a less deeply concave surface for the head of the astragalus and is less extended proximally on the plantar side than in *Felis*. The distal end has three well-defined facets for the cuneiforms.

The *ento-cuneiform* is shaped as in *Felis*, although it supports the hallux, which the modern genus has lost. The *meso-cuneiform* is proportionally smaller than in *Felis* and is of a different shape, narrowing more toward the proximal end. The ecto-cuneiform is the largest of the three, though relatively smaller than that of *Felis*, from which it differs in being narrower and proximo-distally longer. The *cuboid* also is longer and narrower, especially the distal portion, and is therefore of less cubical shape.

The *metatarsals* greatly exceed the metacarpals in length, a disproportion which is greater than in existing cats, though no greater than in *Dinictis*. The hallux, with its phalanges, is slightly longer than the second metatarsal and cannot have been very useful to its possessor. The other metatarsals increase regularly in length from mt. II to mt. IV, but mt. V is a little shorter than II. In *Felis* the metatarsals are nearly parallel and form two symmetrical pairs, II and V shorter, III and IV longer. In *Drepanodon* this arrangement is less distinct, the metatarsals are more divergent and the presence of the hallux gives some appearance of the primitive fan-like disposition of these bones. Mt. II has but a minute lateral contact with the ecto-cuneiform, while in *Felis* this articulation covers half of the proximo-distal length of the tarsal. The third metatarsal is much the stoutest of the series and, except mt. I, V is the most slender; the latter has a protuberance on the fibular side of the proximal end which is very much more prominent than in *Felis*.

The *phalanges* do not exceed those of the manus in length to any such degree as the disproportion in the metapodials would lead one to expect, and are like them in form. The only notable difference is in the unguals, in which the bony hoods are very much less developed, not greatly exceeding the size in *Dinictis*, in which the difference between the ungual phalanges of manus and pes is not so great.

RESTORATION

There is much difference in size, proportions and general aspect of the several species of this genus: *D. oharrai* and *D. occidentalis* are the two largest species, the former characteristic of the Chadron, the latter of the Brulé. Both of these animals were heavy in their proportions, *D. occidentalis*, as above noted, exceeding the Leopard in weight. Of living cats, the nearest approach to these species is to be seen in the Jaguar (*F. onca*). All the species, except, to some extent, the small *D. oreodontis*, have short necks, very long trunk and tail, relatively short limbs and small feet. The primitive dogs, *Daphoenus* and *Pseudocynodictis*, had a similar aspect, except for their long, wolf-like and fox-like heads. These proportions are more viverrine and musteline than like those of modern dogs and cats. Probably, the cat-like viverrine *Cryptoprocta* gives a fairly accurate conception of *Drepanodon primaevus* as a living animal; *D. oharrai* and *D. occidentalis*, as already mentioned, suggest the heavy limbs of the Jaguar. The small *D. oreodontis* differs from the other species, not only in size, but also in its longer legs and more cat-like proportions. Its size is somewhat less than that of the Canada Lynx, though the very long and heavy tail is in striking contrast to that animal.

Species. No less than ten species of this genus have been named, but assuredly this number is too large and should be reduced.

Drepanodon primaevus (Leidy and Owen)

(Pl. XIX)

Machairodus primaevus Leidy and Owen, Proc. Acad. Nat. Sci. Phila., 1851, p. 329.
Drepanodon primaevus Leidy, *ibid.*, 1857, p. 176.
Hoplophoneus primaevus Cope, Bull. U. S. Geol. and Geogr. Surv. Terrs., No. 1, p. 23, 1874.

This is the commonest species of the genus found in the White River formation and is most easily recognized by its moderate size, which is intermediate between the small *D. oreodontis* and the large *D. oharrai* and *D. occidentalis.* Size alone, however, is seldom satisfactory in the definition of a species and *D. primaevus* may best be characterized by its proportions, which are relatively more slender than in the large species and shorter legged and longer faced than in the smaller one. *D. primaevus* has as yet been found only in the Brulé stage of the White River and must be the little-modified descendant of a migrant from Asia and it would seem to have died out without successors, even in the John Day. Several skeletons of this species have been recovered and nearly all parts of its bony structure are well understood. The measurements, for ease of comparison, are combined in the same table with those of *D. oreodontis, D. occidentalis* and *D. oharrai,* the four species which are most distinctly marked as separate. (See p. 139.)

Horizon: Brulé.

Drepanodon oreodontis (Cope)

(Pl. XVI)

Machaerodus oreodontis Cope, Synop. New Vert. Color., Misc. Publ. U. S. Geol. Serv.
Terrs., 1873, p. 9.
Hoplophoneus oreodontis Cope, Ann. Rept. U. S. Geol. Surv. Terrs. for 1872, p. 509, 1873.

The smallest species yet recognized from the White River beds (Brulé substage) though but little smaller than *D. primaevus,* differs in more respects than size from that species. The sabre is shorter and less curved, though this may be a sexual, rather than a specific character. The face is relatively shorter and the upper contour of the cranium shorter and more horizontal. The limbs are proportionally longer and more slender.

MEASUREMENTS (p. 139)

Horizon: Brulé.

Drepanodon molossus (Thorpe)

Hoplophoneus molossus Thorpe, Amer. Journ. Sci. (4), L, p. 220, 1920.

"One very obvious difference between this species and the nearest to which it bears resemblance is the heavy, massive bones which make up all parts of the specimen. The cranial capacity is small and the very small diameter of the postorbital constriction is very noticeable. . . . The face is broad and the cranium short, while the reverse is true of *H. primaevus*" (Thorpe, *loc. cit.*). To this may be added the remarkably anterior position of the postorbital constriction, which is so far forward that it is concealed from view, when the skull is seen from below. In this respect the two extremes within the genus are *D. oharrai,* in which the constriction is as far behind the orbits as in *Hyaenodon,* and *D. molossus,* in which the constriction is as far forward as possible. Postglenoid and post-tympanic processes are very near together. The dimensions following are taken from Thorpe's paper (*loc. cit.*).

MEASUREMENTS

Skull, median basal length........... 132.0 mm.
Skull, width over zygomata (est.)..... 104.0
Skull, width at post-orb. constr....... 26.0
Skull, width over mastoid proc........ 60.0
Mandible, length.................... 109.5

Mandible, length of symphysis........ 39.0 mm.
Mandible, depth of flange (est.)........ 41.0
Tooth-row, length c to m1, incl........ 59.0
Upper canine, ant.-post. diam......... 15.5
Upper canine, transverse diam........ 7.5

Horizon: Brulé.

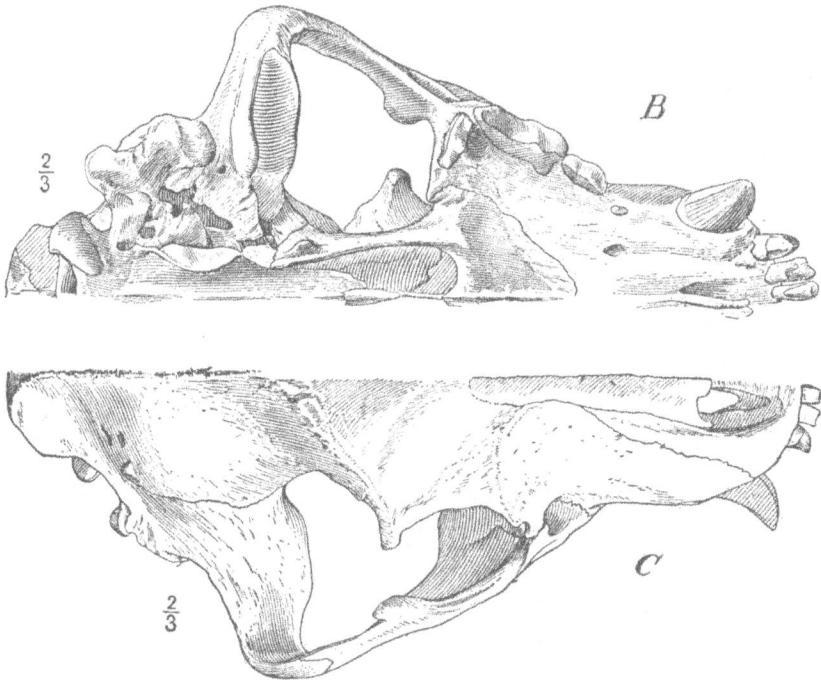

FIG. 7. *Drepanodon molossus;* B, skull from below; C, skull from above. (after Thorpe)

Drepanodon occidentalis Leidy

(Pl. XXII, Figs. 1, 2)

Drepanodon or *Machairodus occidentalis* Leidy, Proc. Acad. Nat. Sci. Phila., 1866, p. 345.
Hoplophoneus robustus Adams, Amer. Nat., XXX, p. 49, 1896.
Hoplophoneus insolens Adams, Amer. Journ. Sci. (4), I, p. 429, 1896.
Dinotomius atrox Williston, Kans. Univ. Quarterly, IX, p. 170, 1895.

The largest and heaviest species of the genus, equalling a modern Leopard in size, though with certain differences of proportions. The upper canines and mandibular flanges are longer and more prominent than in the other species from the Brulé substage, though surpassed in this respect by the species from the Chadron. In some of the skulls referable to this species, the heavy mastoid pedicles are notably smaller and shorter than in others.   Presumably, these are females, though there appears to be no sexual distinction in the length and diameter of the sabres.

MEASUREMENTS (p. 139)

Horizon: Brulé.

Drepanodon oharrai (Jepsen)

(Pl. XV)

Hoplophoneus oharrai Jepsen, Black Hills Engineer, XIV, No. 2, p. 1, 1926.

The most ancient of American sabre-tooth cats as yet made known, is *D. oharrai*, from the Chadron formation at a lower level than *D. mentalis*.   It differs much more decidedly from the Brulé species than the latter do from one another and is, indeed, so distinct that it has been proposed to erect another genus for its reception.   In it are united primitive and specialized characteristics in very confusing fashion.   The dental formula is the same as in the other species and the only peculiarities of the teeth are (1) the diastema between the upper incisors and the canine tusk, which is much longer than in the other species, and (2) the great size of the sabres.   In the type-specimen both tusks were broken off near the base long before the death of the animal, for the stumps are worn smooth and polished by continual abrasion.   This is one of the few cases yet reported of sabres broken off during the lifetime of the animal among the White River machairodonts, in which the great mandibular flanges protected the tusks, but, as previously mentioned, such cases are frequent in the tar-pools of Rancho La Brea.   This, again, is a confirmation of the stabbing hypothesis, for the animal would hardly attempt to bite with the stumps, yet could still deliver formidable blows with the head.

The skull is very elongate and the postorbital constriction is placed far behind the eye-sockets, as in *Hyaenodon;* the contrast between *D. oharrai* and *D. molossus* in this feature is remarkable.   This is one of the long-faced species, space being required for the sheath of the great canine.   The brain-case is smaller than in the species from the Brulé beds, giving a higher sagittal crest, the dorsal border of which is concave.   The par-occipital processes are heavy and project backward from the bullae, with which they are not in contact.   The mastoid pedicle is very large and rugose, as well-developed and conspicu-ous as in the species of this genus from the Brulé substage.   In the type-specimen (So. Dakota School of Mines collection) the auditory bullae are partly preserved in such a manner as to indicate that their major portion was cartilaginous.   The bony part of the bullae shows that they were divided into two chambers, with an arrangement more like that found in the Viverridae than in recent cats, inasmuch as the ento-tympanic seems to have been situated more posterior than median to the ecto-tympanic, which outlines the large external auditory meatus.

The hard palate is relatively longer than in the other White River cats because of the very long muzzle and forward extension of the premaxillaries, for the incisors are not in a straight transverse row, but form a curved line.   The ascending ramus of the premaxilla extends back between maxillary and nasal bones to an unusual distance.

Of particular interest is the series of modifications, which enabled the animal in life to open its jaws to the almost unbelievable angle of 165°–170°.   The shape of the occipital condyles indicates that the head could be thrown back to an unusual degree, thus widening the gape.   The angular process of the mandible is strongly everted, so as to avoid contact with the postglenoid process, when the jaws are opened to the utmost possible extent, a contact which would have limited movement of the mandible.   The great size of the pedicle formed by the united post-tympanic and mastoid processes is characteristic of all advanced machairodonts and is usually proportionate to the dimensions of the canine sabres and is therefore very large in the present species.

The *cranial foramina* differ slightly from those of other species of the genus; the alisphenoid canal and the *foramen ovale*, which are separate in the other species, have a common opening, and there is a foramen, not found in the other Oligocene cats, which is lateral to the *foramen ovale* on the ridge leading to the glenoid fossa.   The mandible has an almost sessile condyle and a remarkably deep flange for the protection of the upper canine, which is relatively deeper than in *Eusmilus;* it is narrower than in other species of *Drepanodon* and, in the type-specimen the free end is pointed, but in the Princeton skeleton it is rounded.

The skeleton (Pl. XV) resembles that of *D. primaevus* and *D. occidentalis* in having a long body and tail, and short, massive limbs, which are proportionately as stout as in the latter species.   One lumbar vertebra shows signs of *spondylitis deformans*, a disease frequently met with in *Smilodon*.

The humerus has a great development of the deltoid ridge, making the proximal two-thirds of the shaft extremely thick antero-posteriorly and the distal end is made very broad transversely by the prominence of the supinator ridge.—Some measurements of the type-specimen are as follows.

MEASUREMENTS

Skull, median basal length	185.0 mm.	P$\underline{3}$ ant.-post. diam.	12.0 mm.
Skull, width over zygomatic arches	120.0	P$\underline{3}$ transverse diam.	5.0
Skull, width of postorb. constr.	27.5	P$\underline{4}$ ant.-post. diam.	19.0
Skull, width over mastoids	77.5	P$\underline{4}$ transverse diam.	8.0
Tooth-row, c to m$\underline{1}$, incl., length	79.0	Mandible, length	150.0
Tooth-row, p$\underline{3}$ to m$\underline{1}$, incl., length	32.0	Mandible, depth of flange	81.0
Diastema before c	8.0	P$\overline{4}$, ant.-post. diam.	11.0
Diastema behind c	18.0	M$\overline{1}$, ant.-post. diam.	17.0
Upper canine, ant.-post. diam.	22.0	Humerus, length	182.0
Upper canine, transverse diam.	9.0	Femur, length	192.0

Horizon: Chadron.

Drepanodon mentalis (Sinclair)

Hoplophoneus mentalis Sinclair, Proc. Amer. Phil. Soc., LX, p. 96, 1921.

This was the first species of the cats to be found in the lower White River, or Chadron substage, and is, so far, represented by a lower jaw (Princeton Mus. No. 12,515).

Sinclair's original description reads: this "is strikingly differentiated from all of the species of the overlying Oreodon beds by the extraordinary depth of the chin." *D. mentalis* is considerably smaller than *D. occidentalis*.   The mandibular flange is U-shaped, with rounded ventral border.

MEASUREMENTS

	D. pri-maevus A.M.N.H., No. 1405	D. occi-dentalis A.M.N.H., No. 1401	D. insolens A.M.N.H., No. 655	D. oreo-dontis Pr. Un. Mus., No. 13,628	D. oharrai Pr. Un. Mus., No. 13,593
	mm.	mm.	mm.	mm.	mm.
Skull, length fr. i1 to occ. cond.............	180.0			149.0	
Cranium, length, ant. rim of orb. to occ. cond.	112.0			99.0	
Face, length, i1 to ant. rim of orbit.........	73.0			55.0	
Skull, width at postorb. constr.............	32.0			27.0	
Skull, width of occiput, at base............	59.0			47.0	
Skull, width over zygomata................	120.0				
Upper canine, ant.-post. diam..............	12.0			11.0	
Mandible, length fr. iĪ to condyle..........	134.0			106.0	
Mandible, length fr. iĪ to angle............	135.0			110.0	
Mandible, height of condyle...............	23.0			18.0	
Mandible, height of coronoid..............	41.0			36.0	
Mandible, depth of flange fr. dors. border....	47.0			57.0	
Atlas, ant.-post. length over cotyles........	34.0				
Atlas, width over trans. proc...............	80.0				
Axis, length of centrum...................	20.0				31.0
Axis, ant.-post. width of neur. spine........	70.0				
Axis, width over transv. processes...........	48.0				
Cervical 4, length of centrum..............	20.0				22.0
Cervical 4, width over transv. proc..........	58.0				
Cervical 6, length of centrum..............	14.0				19.0
Lumbar 1, length of centrum..............	24.0			20.0	26.0
Lumbar 1, width over transv. proc..........	34.0				
Lumbar 3, length of centrum..............	29.0			23.0	28.0
Lumbar 3, width over transv. proc..........	55.0				
Lumbar 6, length of centrum..............	31.0			27.0	34.0
Lumbar 6, width over transv. proc..........	66.0				
Sacrum, length...........................	70.0			60.0	
Caudal 1, length.........................	19.0				22.0
Caudal 1, max. width.....................	56.0				
Caudal 8, length.........................	29.0				32.0
Caudal 13, length........................	26.0				
Scapula, prox.-dist. length.................	137.0			106.0	
Scapula, greatest width...................	89.0				
Scapula, width of acromion................	27.0				
Pelvis, length............................	189.0				191.0
Pelvis, width over ilia....................	103.0				
Pelvis, length of ilium....................	107.0				109.0
Pelvis, length of ischium..................	82.0				82.0
Pelvis, width of iliac plate................	33.0				31.0
Pelvis, length of symphysis................	56.0				58.0
Humerus, length fr. ext. tuberos............	175.0	? 210.0	198.0	142.0	? 174.0
Humerus, width of prox. end..............	35.0	52.0	49.0	30.0	
Humerus, width of dist. end over epicond.....	53.0	? 66.0	62.0	41.0	58.0
Radius, length...........................	133.0	169.0	161.0	111.0	132.0

MEASUREMENTS—*Continued*

	D. pri-maevus A.M.N.H., No. 1405	D. occi-dentalis A.M.N.H., No. 1401	D. insolens A.M.N.H., No. 655	D. oreo-dontis Pr. Un. Mus., No. 13,628	D. oharrai Pr. Un. Mus., No. 13,593
	mm.	mm.	mm.	mm.	mm.
Radius, width of prox. end................	21.0	29.0	26.0	21.0	25.0
Radius, width of dist. end.................	26.0	33.0	27.0	24.0	29.0
Ulna, length fr. olecranon.................	178.0	227.0	209.0		180.0
Ulna, length fr. coronoid proc.............	148.0	182.0	172.0		151.0
Ulna, width at prox. radial facet...........	19.0		22.0	15.0	22.0
Ulna, width of dist. end...................	11.0	12.0	15.0		
Femur, length from gr't troch..............	203.0		231.0	169.0	195.0
Femur, width of prox. end.................	? 48.0		60.0	40.0	51.0
Femur, width of dist. end.................	44.0		50.0	29.0	43.0
Tibia, internal length.....................	163.0		182.0	136.0	151.0
Tibia, width of prox. end.................	42.0		48.0	30.0	43.0
Tibia, width of dist. end.................	27.0		37.0	22.0	25.0
Metacarpal I, length.....................	18.0			14.0	16.0
Metacarpal I, width prox. end.............	10.0			7.0	8.0
Metacarpal II, length....................	24.0	35.0		20.0	28.0
Metacarpal II, width prox. end............	9.0	18.0		7.0	9.0
Metacarpal III, length...................	35.0	49.0		29.0	32.0
Metacarpal III, width of prox. end.........	15.0	18.0		11.0	13.0
Metacarpal IV, length....................	33.0	48.0	47.0	32.0	55.0
Metacarpal V, length....................	30.0 ·	45.0		29.0	26.0
Digit III, phalanx 1, length..............	25.0	(IV) 35.0	(IV) 35.0	20.0	28.0
Digit III, phalanx 1, width prox. end........	18.0	(IV) 19.0	(IV) 17.0	9.0	15.0
Digit III, phalanx 2, length..............	22.0	(IV) 29.0	(IV) 27.0	17.0	21.0
Digit III, phalanx 2, width prox. end........	12.0	(IV) 15.5	(IV) 15.0	10.0	12.0
Digit III, phalanx 3, length..............	21.0			17.0	
Calcaneum, length.......................	51.0		62.0	39.0	54.0
Calcaneum, width over sustent............			29.0	21.0	23.0
Calcaneum, width of distal end............	14.0		17.0	11.0	17.0
Astragalus, length.......................			35.0	25.0	31.0
Astragalus, width of trochlea..............			20.0	15.0	18.0
Metatarsal I, length.....................	25.0			21.0	27.0
Metatarsal II, length....................	39.0			33.0	40.0
Metatarsal II, width prox. end............	11.5			11.0	8.0
Metatarsal III, length...................	40.0		54.0	35.0	33.0
Metatarsal III, width, prox. end...........	17.0		15.0	9.0	12.0
Metatarsal IV, length....................	47.0		59.0	40.0	45.0
Metatarsal IV, width prox. end............	13.0		10.0	7.0	9.0
Metatarsal V, length....................	40.0		48.0	36.0	37.0
Metatarsal V, width prox. end............	9.0		12.0	7.0	9.0
Digit III, phalanx 1, length..............	24.0			20.0	22.0
Digit III, phalanx 1, width prox. end........	11.0			8.0	11.0
Digit III, phalanx 2, length..............	21.0			14.0	15.0
Digit III, phalanx 2, width of prox. end......	10.0			9.0	9.0
Digit III, phalanx 3, length..............				15.0	15.0

Eusmilus Gervais

(Pl. XX)

Machaerodus Filhol, in part, Recherches sur les Phosphorites du Quercy, 1877, p. 153.
Eusmilus Gervais, Zoologie et Paléont. Gen. 2ᵐᵉ sér, 3ᵉ livr., p. 53, 1876.

By far the most remarkably specialized of American Tertiary machairodonts are the two species of this genus, which characterize the lower and upper Brulé substages of the White River Oligocene. Discovered and named in France, it was first recognized in North America by Hatcher, who in 1895 found a very perfect *ramus mandibuli* in the *Protoceras* channel-sandstones of the upper Brulé and named it *Eusmilus dakotensis*. In 1927 Sinclair and Jepsen named a second species, *E. sicarius*, from the lower Brulé, founded upon a fine skull. Another skull of *E. sicarius*, in the collection of the U. S. National Museum, Washington, is of a young animal with unworn teeth and the permanent canines beginning to erupt. Other skulls of *E. dakotensis* are in the museums of the South Dakota State School of Mines and the University of Nebraska. As yet, little of the skeleton, other than skull and mandible, has been obtained.

DENTITION

The dental formula is: $i\frac{3}{2-3}$, $c\frac{1}{1}$, $p\frac{2}{1}$, $m\frac{1}{1}$, though there is some variability in the number of teeth.

Upper Teeth. The *incisors* are simple, sharp-pointed and recurved; i1 and 2 are of nearly equal size and i3 is considerably larger. No example of the immense *canine* has been recovered, though its diameters are plainly given by the empty alveoli and its length may be inferred from that of the protective flange of the lower jaw. Further, the Washington skull of *E. sicarius* retains the greater part of a temporary sabre and from these various indications the great tusk may be confidently restored. No other known sabre-tooth, except the great Pleistocene *Smilodon*, has such canines as those of *Eusmilus*, which are remarkable for their great antero-posterior breadth and for their thinness transversely, the ratio in the type of *H. sicarius* being 32 : 6 (probably too small because of lateral crushing); in the National Museum skull, it is 33 : 8.5, while in one skull of *E. dakotensis* (Rapid City Mus.) it is 46 : 13.7 and the Nebraska University skull of the same species it is 40.5 : 15.3. These ratios are in remarkable contrast to those of *Drepanodon*, in which the thickness is seldom less than one-half of the antero-posterior width.

The *premolars* do not normally exceed two, but a minute, peg-like tooth comes after the canine on one side of the immature Washington skull; it may be a milk-tooth and, at all events, has not been found in any adult jaw. The third premolar (p3) is very small and resembles that of *Drepanodon*, having a central, compressed-conical cusp, with anterior and posterior basal cuspules; in old animals this tooth is much worn on the inner face. The sectorial (p4) is very small, but must have been the only efficient member of the upper cheek-tooth series. When unworn, the crown is a shearing blade of two cusps, of which the anterior is acutely pointed, the posterior a cutting ridge. A minute anterior basal cusp is also present, but is early worn away; there is no internal cusp other than a buttress on the anterior part of the blade, supported by a third root. The only molar (m1) is a small transversely oval tooth, which is not concealed by p4.

Lower Teeth. The anterior part of the mandible between the canines is very narrow so that there are seldom more than two incisors in each ramus, though there appear to be three in Hatcher's type of *E. dakotensis;* the European mandibles all have two. These teeth are small, with simply conical crowns and laterally compressed roots. The *canine,* of which the crown has not yet been recovered, is very small, in correspondence with the great enlargement of the sabres. The cheek-teeth are reduced to two; the single premolar (p4̄) is implanted by two roots and has a tricuspidate crown much larger than that of the third upper premolar, which it resembles in form, having a high, conical middle cusp, with low anterior and posterior basal cusps. The single *molar* (m1̄) is the sectorial, which is quite different in the two species; in the stratigraphically later *E. dakotensis* the sectorial is very feline, the blade composed of two shearing cusps, the inner cusp and heel having completely disappeared. In the earlier *E. sicarius,* the posterior trenchant cusp is higher and more pointed, the heel, though small, is perfectly distinct, and a vestige of the inner cusp remains. Both of these cheek-teeth (p4̄ and m1̄) have a decided backward inclination.

SKULL

The skull abounds in peculiar and remarkable specializations, all of which are plainly conditioned by the great development of the sabres and the manner in which they were used. From this point of view, *Eusmilus* is far more advanced than any other known machairodont of the Oligocene or Miocene. On the other hand, certain primitive features, such as the small brain-case, are combined with these advanced specializations. In the following account the description given by Sinclair and Jepsen (1927) is made use of in somewhat condensed form, though that description applies primarily to the single species, *E. sicarius,* and to a skull which has been somewhat distorted by lateral crushing. Jepsen's subsequent study (1933) of the skulls of both species contained in the museums of Washington, Lincoln, Neb., and Rapid City, S. D., makes possible the elimination of the effects of crushing.

Lateral Aspect. A notable difference from the skull of *Drepanodon* is the relative elongation of the preorbital region of the maxillary, in order to make room for the hypertrophied sabre, as it may fairly be called. At the same time, the tusk is so thin transversely, that it causes no bulging of the alveolar sheath and thus the side of the face is remarkably flat. "The orbits are probably more lateral in aspect than they would ordinarily be, owing to the flattening which the skull has undergone, and their lower margins lack the convexity which is so noticeable in *Hoplophoneus* [*i.e.,* *Drepanodon*], the surface instead sloping upward and inward toward the eye socket as in *Smilodon*" (S. & J., p. 396). The postorbital processes of both frontal and jugal are much better developed than in *Drepanodon,* leaving the orbit less widely open. The process of the jugal is very broad antero-posteriorly. The zygomatic arch is remarkably flat, short and strongly curved upward, but not outward, as it is in almost all other cats; this fore-and-aft straightness of the arch is seen in uncrushed skulls of both species. The pedicle which carries the glenoid cavity is relatively enormous, even more so in *E. dakotensis* than in *E. sicarius,* and brings the cavity down below the level of the lower teeth, making possible an immensely wide opening of the jaws and resembling the structure in *Smilodon.* The mastoid pedicle is also very large, larger and rougher in *E. dakotensis.* In side-view, the cranium differs decidedly in the two species, as will be shown in the specific descriptions.

Dorsal Aspect. Length and narrowness of the skull, the long, straight preorbital muzzle, the small capacity of the brain-case, the great depth of the postorbital constriction and its very posterior position are all conspicuous, when the skull is seen from the top. The postorbital processes of the frontals also display their unusual length in this aspect.

Occipital Aspect. Making all allowance for the effects of lateral crushing, the occiput is narrower at the base and contracts less dorsally than in *Drepanodon* and the condyles are relatively smaller. In this view, there are three pairs of downward projections, which end at different levels. First and hindmost are the paroccipital processes, the distal ends of which are broad; these processes unite suturally with the mastoids, and hence are widely removed from any possible contact with the auditory bullae. Secondly, there are the greatly developed mastoid processes, which are relatively much larger than even in *Drepanodon* and have broad, rugose, freely ending tips. The sterno-cleido-mastoid muscles were attached to these rugose tips and also to large areas of the postero-lateral sides of the mastoids. As Matthew, Merriam and Sinclair have all pointed out, this extraordinary expansion of the areas to which the mastoid muscles were attached, indicates great power in the downward, stabbing stroke of the head. Finally, there are the astonishing glenoid pedicles, which are of relatively enormous height and breadth. Their downward extension brings the joint of the mandible below the level of the lower teeth, facilitating the very wide opening of the mouth. It does not seem possible that so wide a gape could have been made with normally developed temporal muscles, but the bones to which these muscles were attached give evidence that they were weak and lax. The glenoid cavities are very large in both dimensions, but shallow.

Basal Aspect. The undistorted skull in the National Museum enables us to correct the somewhat falsified impression given by the type-skull of *E. sicarius*, due to lateral crushing. Even in the undistorted skull, relative length and narrowness are conspicuous, though the narrowness is less extreme. The premaxillaries form a relatively large portion of the palate and, in the Washington skull they extend well forward of the canines and the alveolar border shows the incisors one behind the other, while in the type-specimen they seem to form a nearly straight, transverse row. There is, however, some doubt about this arrangement. The bony palate is broadest between the premolars, narrowing forward to the minimum width between the canines. The posterior edge of the palate, bounding the narial opening, is a raised border. The narial canal is long, and is broadest in front, contracting posteriorly. The basisphenoid and basioccipital are deeply concave, with elevated lateral borders. No ossified tympanic bulla has been found in connection with any of the skulls and the presumption is that ossification did not take place, though a membranous, or cartilaginous structure must have occupied the open fossa, at the top of which the periotic is exposed.

The *Cranial Foramina* agree with those of the other Oligocene machairodonts, *Dinictis* and *Drepanodon*. An alisphenoid canal is present and opens into a groove common also to the *foramen ovale;* the *foramen rotundum* and *foramen lacerum posterius* are as in the genera named and the carotid canal and condylar foramen are separate from the posterior lacerated foramen.

The *Mandible* is one of the strangest parts of this highly peculiar, not to say grotesque skull. The most immediately striking feature of this jaw is the immense flange, which projects ventrally from the anterior part of each horizontal ramus, and is much larger than in any other known genus of machairodonts. The great "Sabre-tooth Tiger" of the Pleistocene (*Smilodon*) had even larger canine tusks, yet the lower jaw had hardly any protective bony flanges, little more, indeed, than angulations of the ventral border. In *Eusmilus* the flange is broadly concave on the outer side and its lateral and anterior surfaces are demarcated by a sharp ridge. The chin is very deep vertically, because of the great depth of the symphysis, which so unites the two rami that only the ventral tips of the flanges are separate. The depth of the symphysis on the flanges, and especially its median constriction with broadening above and below, seems to be of generic significance, for the three known species of *Eusmilus* all agree in this respect. There is, however, a specific difference in the position of the ventral half of the symphysis, which in *E. sicarius* is in the middle of the chin-flange, while in *E. dakotensis* and the European *E. bidentatus* it is along the anterior border of the chin. Even in the type of *E. sicarius*, which was an old animal with worn teeth, the two mandibular rami are separate, unankylosed at the symphysis.

The horizontal ramus of the jaw is relatively short and weak; below the cheek-teeth it is shallow, but in front of them the depth increases rapidly to the flange. The angular process is very small and is somewhat inflected, so as to be deeply concave on the inner side. The masseteric fossa is deeply impressed, but has no posterior boundary, which is a very unusual feature. The condyle is placed very low down, beneath the level of the alveolar border and the coronoid process is very short, especially in *E. sicarius*, in which it does not rise above the pointed tip of the sectorial. Such a greatly reduced coronoid is indicative of a small, under-developed temporal muscle, an inference confirmed by the low and short sagittal crest and almost undiscernible temporal ridges.

"No other sabre-tooth could equal *Eusmilus sicarius* in ability to open its jaws which, in the absence of muscles, can now be swung backward a full 180° from their position of complete closure before the condyle begins to rise out of its socket. This is possible because of the great length of the glenoid process, and the relatively slight elevation of its posterior inner corner, the shallow glenoid fossa, the pedunculate condyle and the small angular process which does not interfere with the rear of the glenoid process until the full 180° maximum of gape is attained" (Sinclair and Jepsen, p. 405).

Species. Two well-distinguished species of this genus, *E. sicarius* and *E. dakotensis*, characterise respectively the lower and upper Brulé, and there is no reason to separate these generically from the European *E. bidentatus*. The latter, however, is known only from the lower jaw and if the skull were available, it might prove necessary to assign the American species to another genus, an assignment which would simplify the problems concerning the origin and relationships of *Eusmilus*. These problems will be considered in a following section.

Eusmilus sicarius Sinclair and Jepsen

(Pl. XX)

Eusmilus sicarius Sinclair and Jepsen, Proc. Amer. Phil. Soc., LXVI, pp. 391–407, 1927.

In addition to the type specimen (P. U. Mus., No. 12,953) which is a fine skull, with mandible, somewhat distorted by crushing, there is a referred skull, without lower jaw, in the U. S. Nat. Mus. (No. 12,820) which is that of a young animal with milk-canines in place and the permanent ones just beginning to erupt.

E. sicarius is smaller than *E. dakotensis* and has a less specialized lower sectorial, retaining a distinct trenchant heel and a vestigial paraconid. The mastoid pedicle is not so greatly enlarged and the sagittal crest is not so straight or horizontal, rising to the occipital crest, so as to give the hinder part a concave margin. The angular process of the lower jaw is much smaller and the coronoid lower.

MEASUREMENTS

	P. U. *No. 12,953*	Nat. Mus. *No. 12,820*
Skull, length med. bas. line	187.0 mm.	211.0a mm.
Skull, length prmx. to occ. crest	198.5a	218.5
Skull, length prmx. to bregma	142.0a	165.0
Skull, length prmx. to post. edge of can. alveo.	48.0a	58.5
Skull, length prmx. to p3	69.0a	81.0a
Upper canine alv. ant.-post. diam.	32.0	33.0
Upper canine alv. transverse diam.	6.0	8.5a
P3 to m1, length	36.0	41.8
P3 ant.-post. diam.	9.0	11.4
P4 ant.-post. diam.	22.5	22.0
Palatal length	115.0a	120.0
Mandibular ramus, length	167.0	
Mandible, depth of symphysis	84.0	
Mandible, depth behind p4	33.0	
Mandible, depth of flange	91.0a	
P4–m1, length on alveo. border	31.0	
M1, ant.-post. diam.	19.5	
M1, transverse diam.	8.5	

(a. approximate)

Horizon: Lower Brulé.

Eusmilus dakotensis Hatcher

Eusmilus dakotensis Hatcher, Amer. Nat., XXIX, p. 1091, 1895.

The type of this species is a complete half of a lower jaw, with canine broken and i1 missing. Jepsen (1933) has assigned as an allotype a specimen in the museum of the So. Dak. State School of Mines (No. 2815), a fine skull, without mandible, from which all the teeth except the left third incisor and the right third premolar, have been lost. Another referred specimen is in the Nebraska State Museum (No. 6-12-7-95) and consists of the left half of a skull without lower jaw. *E. dakotensis* differs from its predecessor, *E. sicarius*, in the further development, not to say exaggeration, of the structural features characteristic of that species:

1. It is of larger size, though, in this respect, both species display considerable variability.

2. The mastoid and glenoid pedicles are still larger in proportion to the size of the skull.

3. The upper contour of the sagittal crest is nearly straight and horizontal, not concave.

4. The occiput is higher.

5. The zygomatic arch is even shorter, straighter and more massive.

6. The canine sabre is larger and the lower sectorial (m1) is simplified by the loss of the heel and the vestigial paraconid.

This is the largest of the Oligocene cats and is not equalled by any yet found in the Miocene.

MEASUREMENTS

	Princeton U. Type, No. 11,079	Neb. St. Mus. No. 6–12–7–95	S. D. Sch. M. No. 2815
	mm.	mm.	mm.
Skull, med. basal length................................			251.0
Skull, length prmx. to inion.............................			294.5
Skull, length prmx. to bregma..........................			208.5
Length, prmx. to post. edge of canine alveo..............		61.0a	70.7
Length, prmx. to ant. edge of p3........................		81.0a	86.5
Upper canine, ant.-post. diam...........................		40.5	46.0
Upper canine, transverse diam...........................		13.1	13.7
P3–m1, length..		47.0	43.5
P3, ant.-post. diam. alv................................		12.6	12.6
P4, ant.-post. diam. alv................................		29.6	32.6a
Palatal length (after Eaton)............................		124.0a	129.0
Cranium, width over postorb. proc.......................		86.0	91.0
Cranium, width at postorb. constr.......................			41.0
Mandible, length, inc. alv. to angle incl.................	173.0		
Flange, depth fr. post. edge of c̄........................	79.0		
Diastema, c̄–p4̄...	54.0		
Horiz. ramus, depth at p4̄...............................	36.0		
Horiz. ramus, depth behind m1̄..........................	31.5		
P4̄, ant.-post. diam.....................................	16.0		
M1̄, ant.-post. diam....................................	23.0		
Coronoid, height above line fr. cond. to base of m1̄.........	24.0		

(a. approximate)

Horizon: Upper Brulé.

The problem concerning the origin of *Eusmilus* is obscured by the occurrence of the genus in the Old World. Were it not for that, there would be small difficulty in deriving it from the lower White River species, *Drepanodon oharrai*, though, perhaps, the time interval would be too short. There are several alternative possibilities concerning the origin of the American species of *Eusmilus:* (1) that the older one, *E. sicarius*, was an immigrant, in which case the resemblance to *D. oharrai* is accidental; (2) that *Eusmilus* is of American origin and the French species an immigrant; and (3) that the genus arose twice independently and if such a dual origin of a mammalian genus ever actually took place, the diphyletic origin of *Eusmilus* would not be incredible, for it implies merely the exaggeration of prevailing tendencies. On the whole, however, the first alternative, that *Eusmilus*, like all the other cat genera, was an immigrant, seems the most likely.

Subfamily NIMRAVINAE Subfam. Nov.

Cope included all of the sabre-tooth cats in his family of Nimravidae, a term which is not admissible on grounds of priority. He also made the distinction of the genera *Archaelurus* and *Nimravus*, as the "false sabre-tooths," though without proposing a formal subdivision to receive them. That is here done, as a tentative arrangement, to help in clearing up the very obscure problems which concern the origin and inter-relationships of the cats and the more or less cat-like Viverridae and Hyaenidae, on the one hand and the machairodonts and "false sabre-tooths" on the other.   The latter are, in some degree, intermediate between the true cats and the machairodonts and hence the divergent views which have included them in each of the other subfamilies.

The definition of the Nimravinae differs somewhat according to the inclusion or exclusion of the Miocene and Pliocene genera, *Pseudaelurus* and *Metailurus*.   As only the Oligocene genus is dealt with here, the later genera may be omitted from consideration. The subfamily agrees with the machairodonts in the structure of the cranial basis, in the number and position of the cranial foramina and in the character of the auditory bullae, so far as that has been ascertained.   A tympanic ring of varying size is always present and, when ossified, the bulla is of extremely thin and paper-like bone and is far removed from the paroccipital and mastoid processes.   The mandible agrees with that of the machairodonts in having the chin demarcated from the outer side of the horizontal ramus by a strong angulation.

Differing from the sabre-tooths are the following features: The functional premolars are much larger and higher and better adapted for seizing and holding.   There is no mastoid pedicle and no zygomatic pedicle of the squamosal for the glenoid cavity and no modification of skull-structure to permit an extraordinarily wide opening of the mouth.   The lower jaw is without flanges.   The limb-bones are long and slender, as in the Cheeta (*Cynaelurus*).

Nimravus Cope

(Pl. XXI)

Nimravus Cope, Proc. Acad. Nat. Sci. Phila., 1879, p. 169.

The definition of the subfamily includes *Archaelurus*, which is known only from the John Day and is often merged with *Nimravus*, though, in our judgment, this is not advisable. In *Nimravus* the upper canine is longer, more compressed laterally and with serrate edges; lower canine relatively smaller.   Lower sectorial more cat-like, with posterior cusp of trenchant blade thin and compressed, but separated by a deep notch from the anterior cusp, though the notch is narrower than in *Archaelurus*.

The dental formula is: $i\frac{3}{3}$, $c\frac{1}{1}$, $p\frac{4-3}{3-2}$, $m\frac{1}{2}$.   In both jaws the number of vestigial premolars is variable.

Nimravus bumpensis sp. nov.

To the kindness of Mr. James D. Bump of the State School of Mines of South Dakota, in Rapid City, we owe the opportunity of describing and figuring this fine skull and are glad to name the species in his honour.   The genus has not been previously reported from the White River, though the late Mr. O. P. Hay referred the *Dinictis major* of Lucas to *Nimravus* (2nd Bibliog. and Catal. of Foss. Vert. of N. A., II, p. 543), but this in untenable.

The new species, *N. bumpensis*, is somewhat smaller than the type species, *N. gomphodus* and about equal to *N. confertus*, both of the John Day; it differs from both of the species just named in having an additional premolar in each jaw, thus giving ($p\frac{4}{3}$), but we do not regard the presence or absence of these minute, vestigial teeth as being of generic importance, especially as there are but three upper premolars on the right side; it can be given specific value only if the teeth are constantly present, or absent.   Otherwise, the teeth differ very little from those of the John Day forms; there is a similar increase in the size and especially the height of the functional premolars other than the upper sectorial, p3 and 3 and 4, to which J. C. Merriam has already called attention in both of the Oregon genera.   The incisors are small and decidedly more reduced than in *N. gomphodus* (they are not known in *N. confertus*) and the canines are proportionately somewhat shorter and thicker.   The superior sectorial (p4) has a very prominent inner cusp, or deuterocone, and in the inferior sectorial the notch which separates the two cusps of the trenchant blade is wider than in the Oregon species and almost as wide as in *Archaelurus*.

In the skull there are several differences from *N. gomphodus* (that of *N. confertus* is not known).   The suborbital portion of the maxillary is much narrower, bringing the infraorbital foramen down to a very inferior position.   The zygomatic arches curve out much less boldly from the sides of the skull, narrowing the temporal openings conspicuously. The occiput is of a different shape, with the dorsal crest of the inion less curved and more nearly straight transversely and the paroccipital processes are decidedly shorter and point more directly backward.   The auditory bullae are small and seem to have been ossified only in part; on each side there is a tympanic ring, with large *meatus auditorius*, which ends in a thin plate ventrally.   On each side of the basioccipital the margin is flared downward in an unusually large process which abuts against the bulla.   The angular process of the mandible is shorter.

Of particular interest is a fearful, gaping wound which penetrates the left frontal bone (Pl. XXI, Fig. 1b), piercing to the sinus; that this was inflicted during the lifetime of the animal is shown by the callus along the margins.   Much the most likely explanation is that the wound was given by a stabbing blow from the sabre of *Eusmilus*.

This is really a remarkable case, which strongly supports the stabbing hypothesis concerning the manner in which the great tusks of the sabre-tooth cats were used, for the wound is of the proper size and shape to have been inflicted in the way supposed.   The skull is too narrow to have allowed both canines of the attacking *Eusmilus* to strike it. It may seem surprising that not more instances of such wounds should have been discovered, but there are two reasons for the rarity of them: (1) In the first place, it is probable that the machairodonts struck at the neck and shoulders of their victims, rather than at their heads, for this is the mode of attack followed by the great existing cats.   (2) The partially eaten carcasses left by the sabre-tooth cats, which were manifestly unable to crush bones, were almost certainly disposed of by the scavengers, especially by the species of *Hyaenodon*, the massive premolars of which were well adapted for bone-crushing.   The stabbing hypothesis, unlikely as it must seem, is the only one yet suggested that explains the structural facts of the enigmatical sabre-tooths.

PRINCETON, N. J.,
    April 20, 1936.

MEASUREMENTS

	N. gomphodus	N. confertus	N. bumpensis
	mm.	mm.	mm.
Upper canine, length..............................	45.0		30.0
Upper canine, ant.-post. diam........................	16.0		15.0
Upper canine, transverse diam.......................	8.0		8.0
Upper canine to p3, distance.........................	16.0		15.0
Upper cheek-teeth, p3–m1, (incl.), length...............	45.0		44.0
P3, ant.-post. diam................................	18.0		19.0
P3, height of apex.................................	13.0		12.0
P4, ant.-post. diam................................	25.0		22.0
P4, height of apex.................................	15.0		15.0
M1, width..	9.0		9.0
Upper incisor series, width (i3 to i1).................	18.0		21.0
Lower canine, ant.-post. diam........................	12.0	10.0	10.0
Lower canine, transverse diam.......................			7.0
Lower canine, length...............................	24.0	16.0	20.0
Lower cheek-teeth, p2–m2 (incl.), length...............	63.0	53.0	55.0
P3, ant.-post. diam................................	17.5	14.0	14.0
P3, height of apex.................................	17.5	10.0	12.0
P4, ant.-post. diam................................	20.0	16.0	16.0
P4, height of apex.................................	15.0	13.0	12.0
M1, ant.-post. diam................................	25.0	22.0	22.0
M1, height of apex.................................	16.0	15.0	15.0
Skull, length occ. cond. to prmx. border..............	206.0		178.0
Skull, length inion to prmx. border...................	220.0		191.0
Skull, length of nasal bone..........................	65.0		57.0
Skull, length sagitt. crest..........................	82.0		59.0
Skull, width of prmx. bone..........................	19.0		14.0
Width of nasal bone at middle........................	8.0		7.0
Width of frontal at middle of orbit...................	28.0		23.0
Width of frontal at postorb. proc....................	35.0		34.0
Skull, width at ant. part of zygoma..................	98.0		67.0
Skull, width over zygomatic arches...................	111.0		87.0
Skull, width at meat. audit.........................	74.0		44.0
Skull, width at parocc. proc........................	54.0		41.0
Occiput, width at middle............................	44.0		36.0
Occiput, height above for. mag......................	36.0		30.0
Mandible, width of chin at base......................	22.0		23.0
Mandible, depth of chin.............................	40.0	27.0	33.0
Mandible, depth at diast...........................	27.0	20.0	20.0
Mandible, depth below p4............................	31.0		23.0
Mandible, length...................................	157.0		132.0
Mandible, height of condyle..........................	33.0		30.0
Mandible, height of coron. proc......................	71.0		62.0

N.B. The measurements of the two John Day species are taken from Cope: *Tert. Vert.*, pp. 968 and 973.

Horizon: Nimravus bumpensis has been found only in the highest levels of the Brulé substage of the White River.

EXPLANATIONS OF THE PLATES

PLATE I

Apternodus gregoryi, Mus. Comp. Zool. Harvard Univ.

Fig. 1. Skull, side-view.
Fig. 1a. Skull, base.
Fig. 1b. Skull, top.
Fig. 1c. Skull, occiput.
Fig. 1d. Lower jaw, crown view.
    All figures × 3.

PLATE II

Fig. 1. *Micropternodus borealis*, type, A.M.N.H., lower jaw, × 3.
Fig. 2. *Proterix loomisi*, skull, top view × 3, A.M.N.H.
Fig. 2a. *Proterix loomisi*, skull, base (after Matthew), × 3.
Fig. 3. *Clinodon gracilis* type, lower jaw, side view, × 3, Princeton Mus.
Fig. 3a. *Clinodon gracilis* type, lower jaw, crown view, × 3, Princeton Mus.
Fig. 4. *Proscalops miocaenus*, type; skull, side view, × 2 (after Matthew).
Fig. 4a. *Proscalops miocaenus*, type; skull, top, × 2 (after Matthew).
Fig. 5. *Metacodon magnus* type, lower jaw, side, × 3¼, Princeton Mus.
Fig. 5a. *Metacodon magnus*, type, lower jaw, crown view, × 3¼, Princeton Mus.
Fig. 6. *Daphoenus hartshornianus*, base of cranium, × 1½, Princeton Mus.

PLATE III

Ictops

Fig. 1. Skull, side view, × 3, University of Nebraska, So. Dak. School of Mines.
Fig. 1a. Skull, top, × 3, University of Nebraska, So. Dak. School of Mines.
Fig. 1b. Upper teeth, crown view, Princeton Mus. No. 10,017. Figures slightly less than 3 times nat. size.
Fig. 1c. Upper teeth, side view × 3 Princeton Mus. No. 10,539.
Fig. 2. Manubrium sterni, side view × 3 Princeton Museum.
Fig. 2a. Manubrium sterni, ventral view.

PLATE IV

Fig. 1. *Leptictis haydeni*, type, skull, top, slightly less than × 2.
Fig. 1a. *Leptictis haydeni*, type, skull, side, slightly less than × 2.
Fig. 1b. *Leptictis haydeni*, type, skull, base, slightly less than × 2.
Fig. 1c. *Leptictis haydeni*, type, upper teeth, crown view, × 3½.
    Academy of Natural Sciences, Philadelphia.
Fig. 2. *Ictops dakotensis*, upper teeth, side view, × 3½, Princeton Mus. No. 10,539.
Fig. 2a. *Ictops dakotensis*, upper teeth, crown view, × 3½, Princeton Mus. No. 10,017.

PLATE V

Sinclairella dakotensis, type

Fig. 1. Skull, side view, Princeton Mus. No. 13,585.
Fig. 1a. Mandible, crown view, Princeton Mus. No. 13,585.
    Figures × 3½.

PLATE VI

Sinclairella dakotensis, type

Fig. 1. Skull, top, × 3½, Princeton Mus. No. 13,585.
Fig. 1a. Skull, base, × 3½, Princeton Mus. No. 13,585.

PLATE VII

Hyaenodon horridus, skeleton, A.M.N.H. No. 1375.   1/5 nat. size

PLATE VIII

Hyaenodon mustelinus, Princeton Mus. No. 13,583

Fig. 1. Skull, left side, × 1.
Fig. 1a. Skull, base, × 1.
Fig. 1b. Skull, occiput, × 1.
Fig. 1c. Mandible, left side, × 1.
Fig. 1d. Lower teeth, crown view, × 1.
Fig. 2. Scapula, right side, × 1.   (No. 13,603)
Fig. 3. *H. cruentus*, Princeton Mus. No. 12,580.   Tympanic bone, × 4/5.

PLATE IX

Figs. 1–5. *Hyaenodon cruentus*, Princeton Mus. No. 10,995.
Figs. 6–11. *H. mustelinus*, Princeton Mus. No. 13,603.
Fig. 1. Sixth lumbar vertebra, left side, × 11/9.
Fig. 2. Left scapula, × 3/4.
Fig. 3. Left humerus, anterior side, × 3/4.
Fig. 4. Left femur, anterior side, × 3/4 apprmx.
Fig. 5. Left tibia, anterior side, × 3/4 apprmx.
Fig. 6. Atlas, right side, × 6/5.
Fig. 7. Axis, right side, × 4/3.
Fig. 8. Fifth lumbar vertebra, right side, × 3/2.
Fig. 9. Right humerus, ant. side, × 3/4.
Fig. 10. Right os innomination, × 8/11.
Fig. 11. Right femur, ant. side, × 5/6.

PLATE X

Daphoenus vetus

Fig. 1. Skull, left side, Field Mus., Chicago, No. 12,021.
Fig. 1a. Skull, top, Princeton Mus., No. 12,648.
Fig. 1b. Skull, base, Princeton Mus., No. 12,648.
Fig. 2. Brain-cast, Princeton Mus., No. 12,588.
Fig. 3. Mandible, side, Princeton Mus., No. 13,600.
Fig. 3a. Lower teeth, crown, Princeton Mus., No. 13,600.
      All figures × 2/3.

PLATE XI

Figs. 1–9. *Daphoenus vetus.*
Fig. 10. *Pseudocynodictis gregarius.*
Fig. 1. Right humerus, from behind.
Fig. 2. Right ulna, anterior side.
Fig. 3. Right radius, anterior side.
Fig. 3a. Right radius, distal end.
Fig. 4. Right femur, anterior side.

Fig. 5. Right tibia, anterior side.
Fig. 6. Right fibula, outer side.
Fig. 7. Right pes, dorsal side.
Fig. 8. Right manus, dorsum.
Fig. 9. Baculum, from left side.
Fig. 9a. Baculum, from above.
Fig. 10. Baculum, from right side.
Fig. 10a. Baculum, from above.

These figures redrawn from Hatcher's plates, corrected by the original specimens in the Carnegie Museum, Pittsburgh.

All figures × 3/4.

PLATE XII

Fig. 1. *Daphoenus hartshornianus.* Skull, left side, Princeton Mus. No. 12,956, incisors and some premolars supplied from Nos. 12,650 and 11,421.
Fig. 1a. The same, top view. Figs. 1, 2 × 3/4.
Fig. 1b. The same, base.
Fig. 1c. The same, occiput.
Fig. 2. *Daphoenus vetus,* occiput.
Fig. 3. *Daphoenus dodgei,* type; mandible, right side, × 12/11.
Fig. 3a. *Daphoenus dodgei,* type; lower teeth, crown view, × 12/11.

PLATE XIII

Pseudocynodictis gregarius

Fig. 1. Skull, right side, Peabody Mus. Yale Univ. No. 10,068.
Fig. 1a. The same, base.
Fig. 2. The same, top; brain-cast of No. 10,067 inserted.
Fig. 3. Manus, Princeton Mus. No. 11,012.
Fig. 4. The same, pes.

Figures × 6/4.

PLATE XIV

Fig. 1. *Parictis dakotensis,* mandible, right side, × 7/4, Mus. So. Dak. State Sch. Mines.
Fig. 1a. *Parictis dakotensis,* lower teeth, crown-view, × 7/4.
Fig. 2. *Mustelavus,* type × 10/7, skull, base, Princeton Mus. No. 13,775.
Fig. 2a. *Mustelavus,* type, left upper teeth, × 10/7.
Fig. 2b. *Mustelavus,* type, left lower teeth, × 10/7.
Fig. 3. *Bunaelurus lagophagus,* skull, right side, Princeton Mus. No. 13,588.
Fig. 3a. *Bunaelurus lagophagus,* skull, base.
Fig. 3b. *Bunaelurus lagophagus,* skull, top.
Fig. 3c. *Bunaelurus lagophagus,* skull, occiput.
Fig. 3d. *Bunaelurus lagophagus,* lower teeth, crown view.

Figs. 3–3d × 15/7.

PLATE XV

Upper figure: *Dinictis felina;* skeleton, Mus. So. Dak. State School of Mines, Rapid City, × 1/6 approx.
Lower figure: *Drepanodon oharrai;* skeleton, Princeton Mus. No. 13,593 × 1/5 approx.

PLATE XVI

Upper figure: *Hyaenodon mustelinus;* skeleton, × 1/4, Princeton Mus. No. 13,603.
Lower figure: *Drepanodon oreodontis;* skeleton, Princeton Mus. No. 13,628.

PLATE XVII

Dinictis squalidens

Fig. 1. Skull, right side, Princeton Mus. No. 13,587, × 3/2.
Fig. 1a. Lower jaw, crown view.

PLATE XVIII

Dinictis squalidens, Princeton Mus. No. 13,587

Fig. 1. Skull, top view, × 3/2.
Fig. 1a. Skull, base.

PLATE XIX

Drepanodon primaevus, Princeton Mus. No. 10,540

Skull, right side, × 1/1.

PLATE XX

Eusmilus sicarius. Type, Princeton Mus. No. 12,953

Skull, right side, × 1/1.

PLATE XXI

Nimravus bumpensis. Type × 1/1, So. Dak. State School of Mines, Rapid City

Fig. 1. Skull, left side.
Fig. 1a. Skull, base.
Fig. 1b. Skull, top.
Fig. 1c. Left incisors and canines, front view.
Fig. 1d. Lower teeth, crown view.
Fig. 1e. Lower teeth, inner side.

PLATE XXII

Fig. 1. *Drepanodon occidentalis:* Left manus, Princeton Mus. No. 11,022.
Fig. 2. *Drepanodon occidentalis:* Left pes, Princeton Mus. No. 11,022.
Fig. 3. *Dinictis felina:* Left manus, A.M.N.H.
Fig. 4. *Dinictis felina:* Right pes, A.M.N.H.
Fig. 5. *Hyaenodon cruentus:* Left manus, A.M.N.H.
Fig. 6. *Hyaenodon cruentus:* Right pes, Princeton Mus. No. 10,916.

All figures approx. 3/4 nat. size.

PLATE I

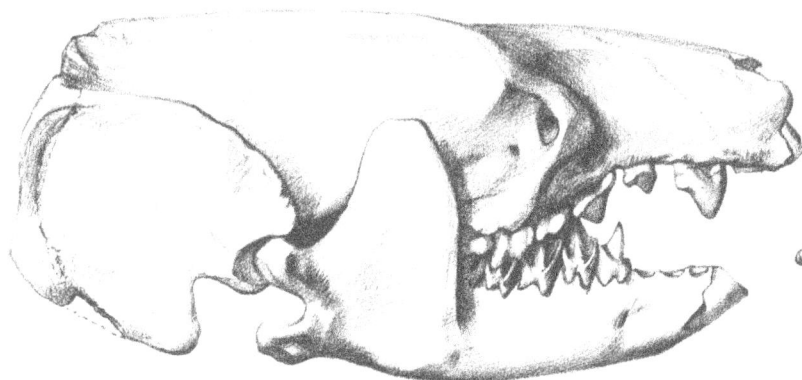

1.

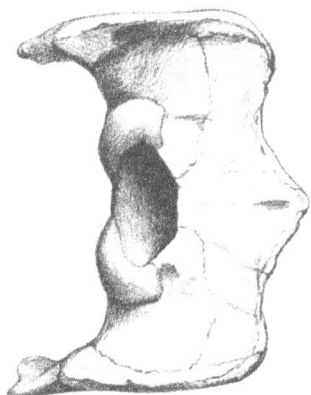

1$\underline{c}$

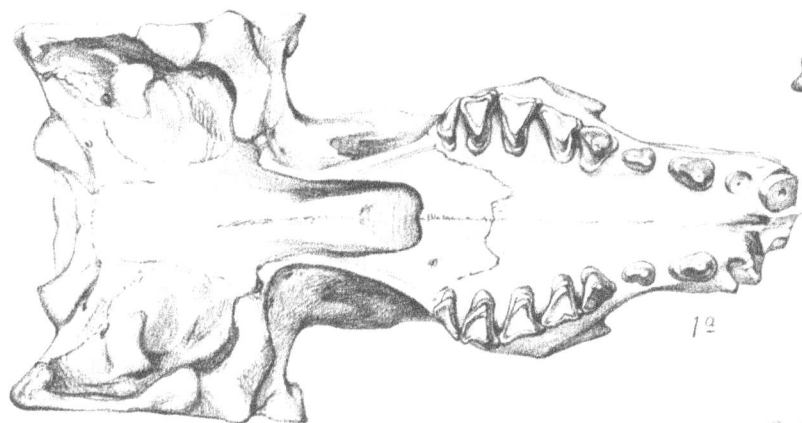

1$\underline{a}$

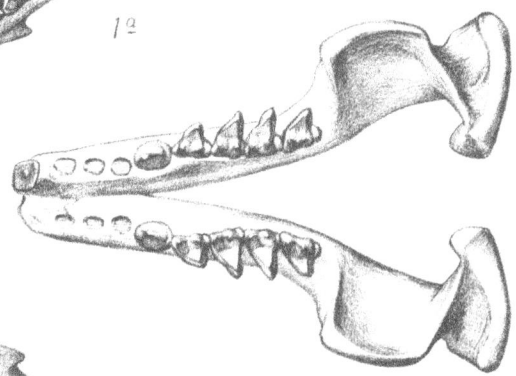

1^d

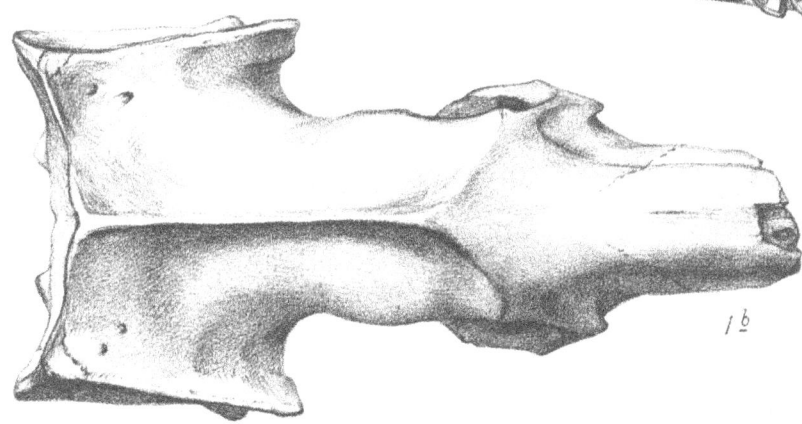

1$\underline{b}$

APTERNODUS

PLATE II

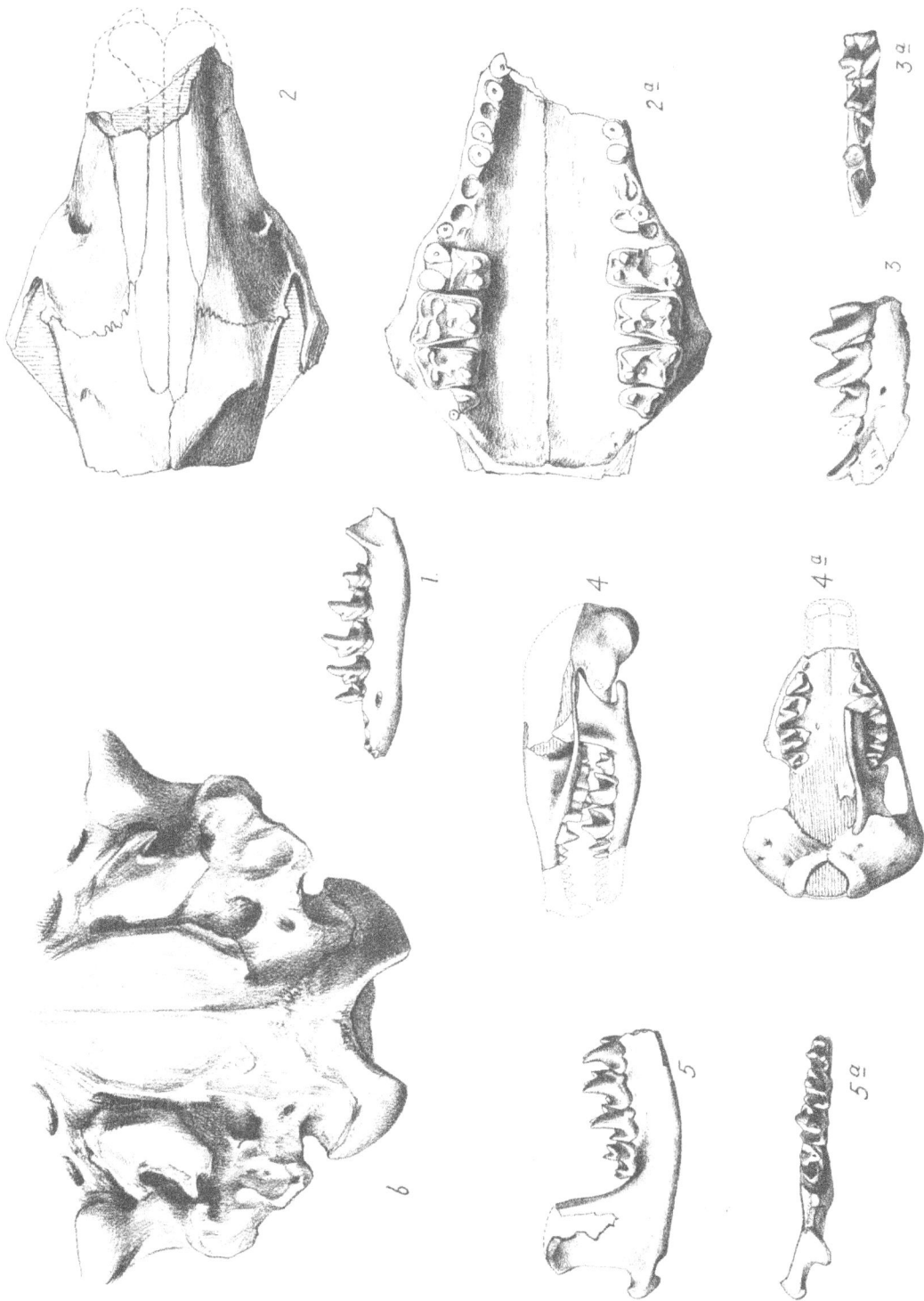

INSECTIVORA—DAPHOENUS

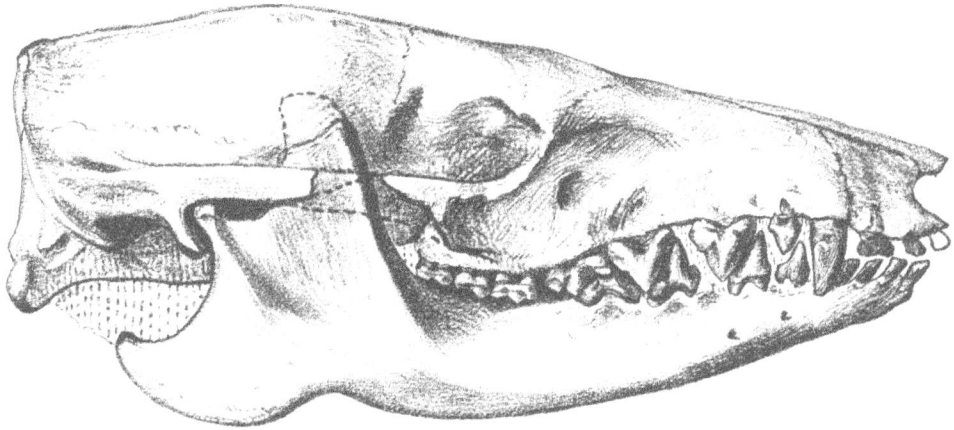

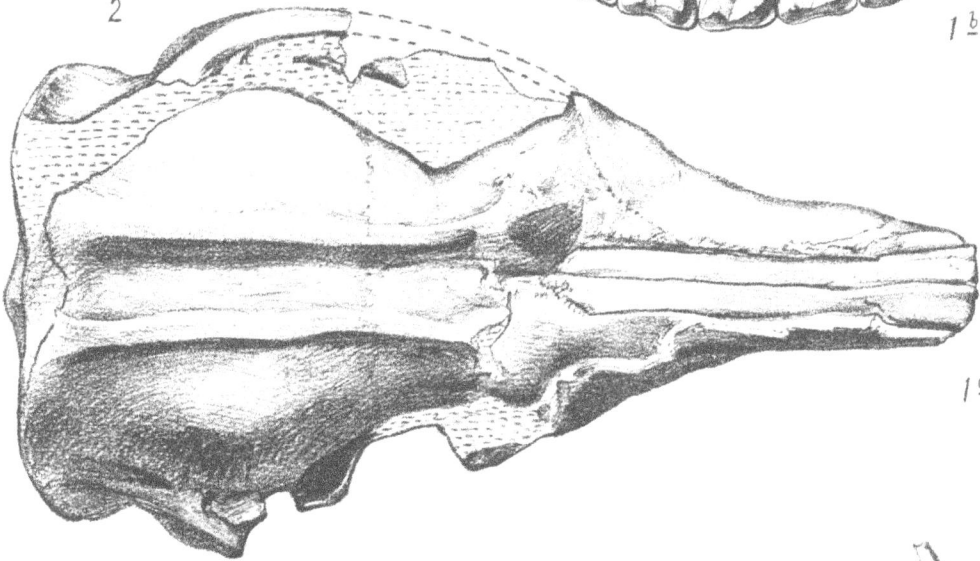

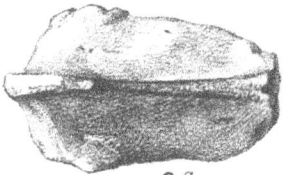

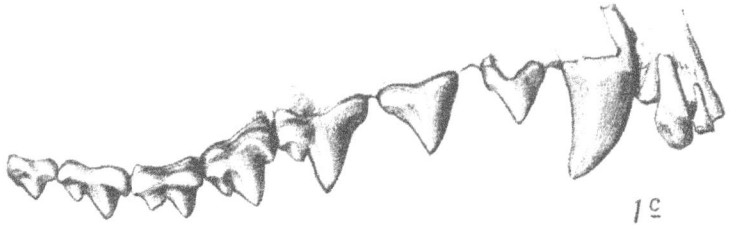

PLATE III

1

2

1 $\underline{b}$

1 $\underline{a}$

2 $\underline{a}$

1 $\underline{c}$

ICTOPS

PLATE IV

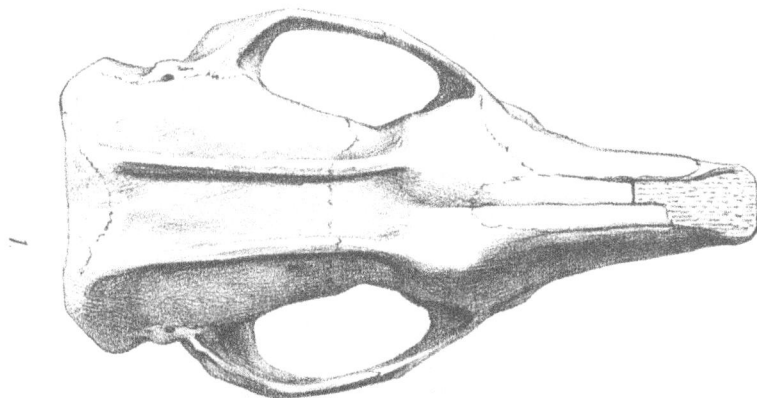

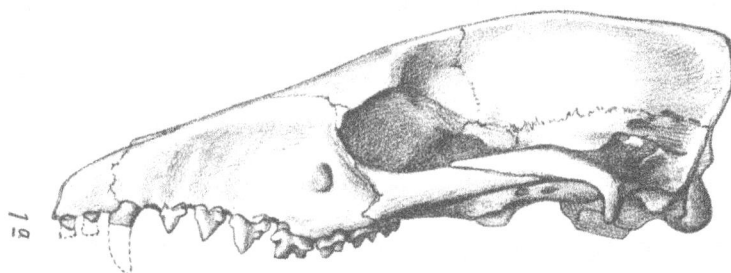

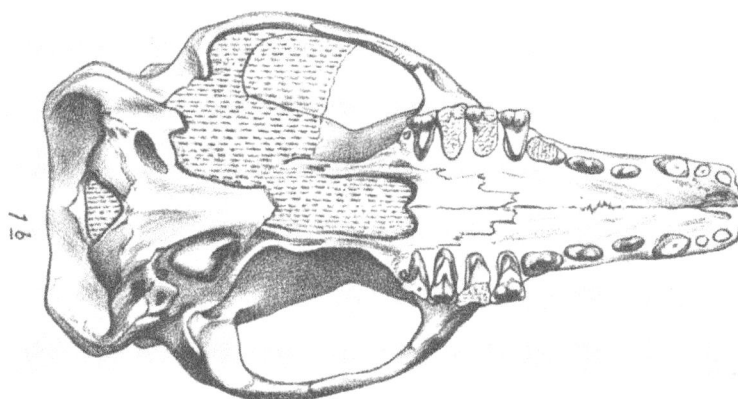

LEPTICTIS—ICTOPS

PLATE V

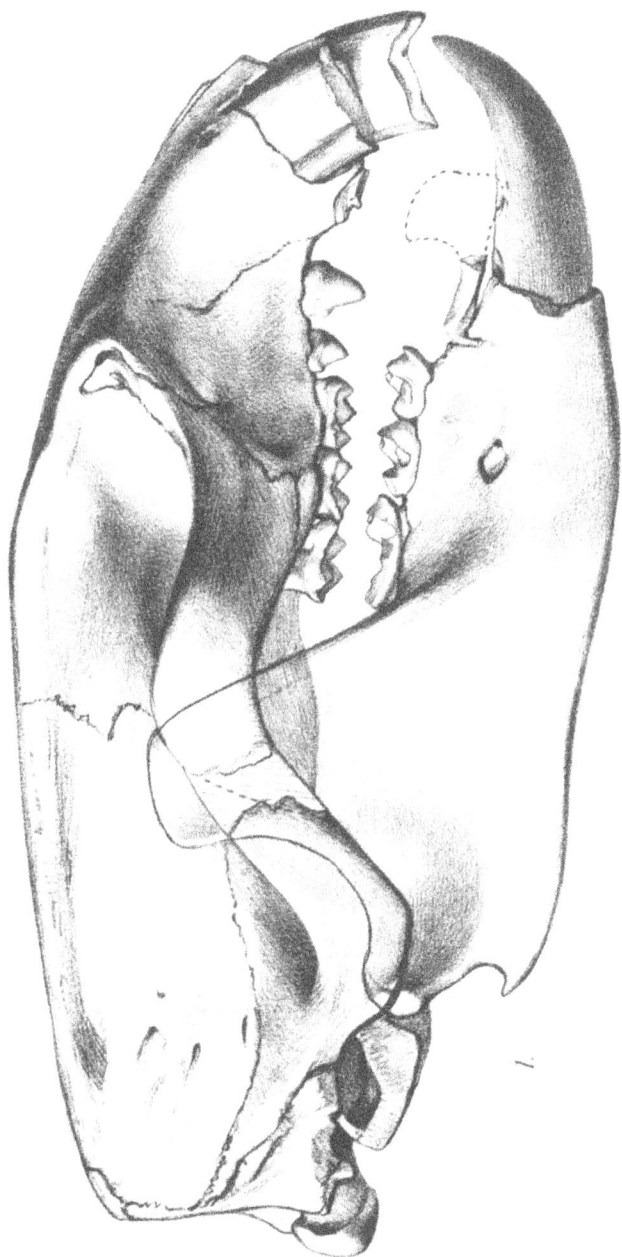

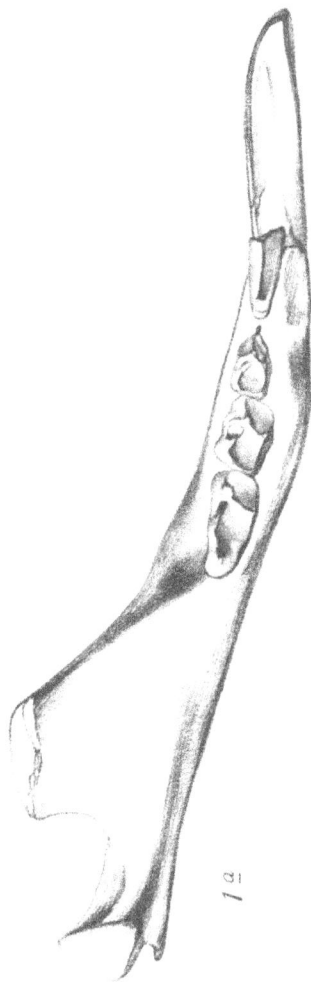

SINCLAIRELLA

PLATE VI

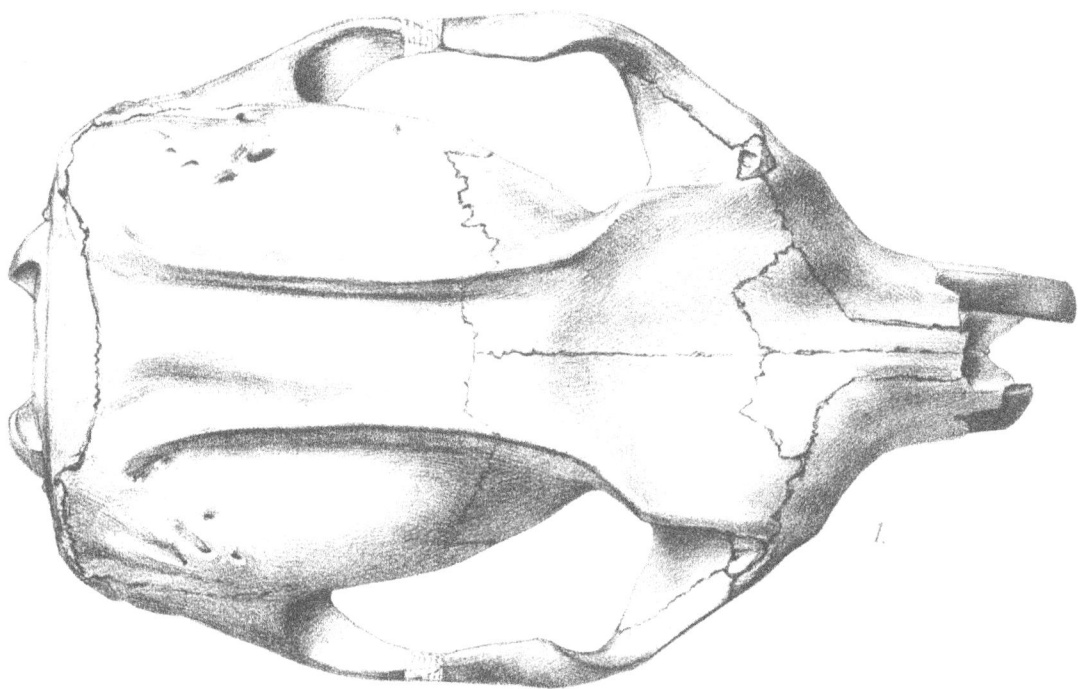

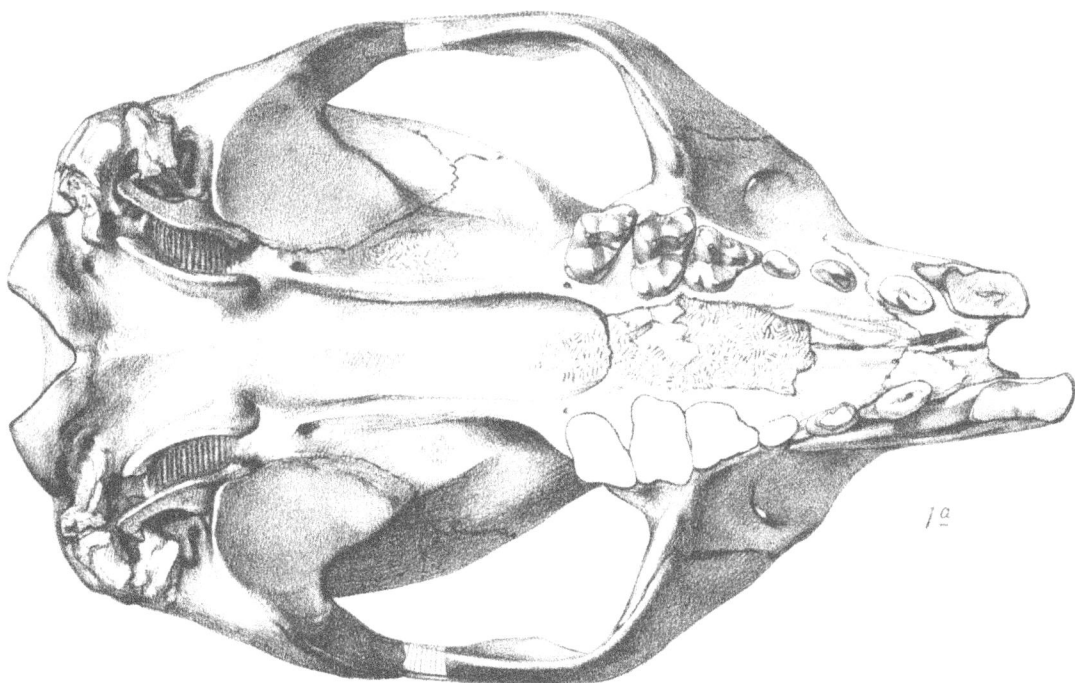

PLATE VII

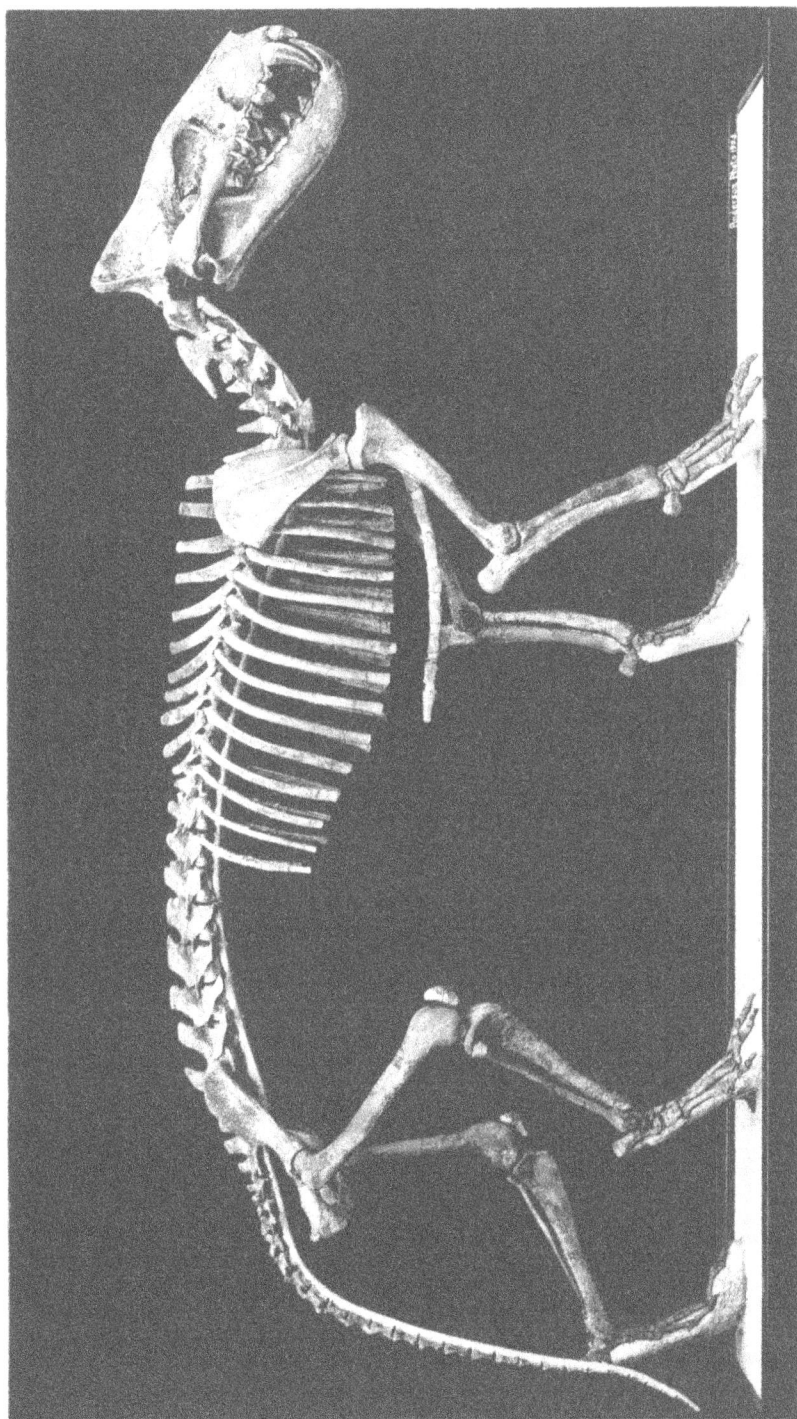

HYAENODON

PLATE VIII

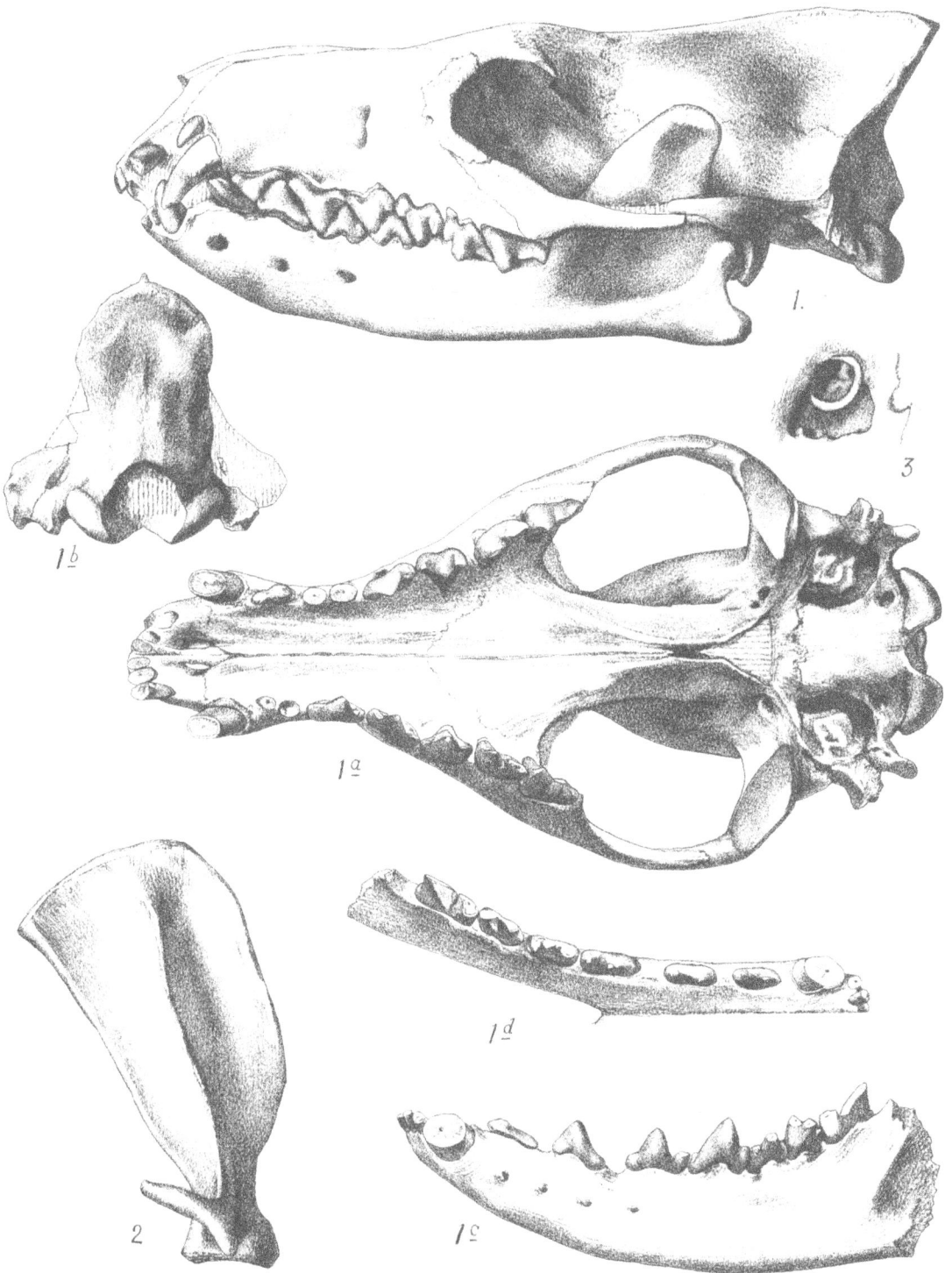

1.

1$\underline{b}$

1$\underline{a}$

3

2

1$\underline{d}$

1$\underline{c}$

HYAENODON

PLATE IX

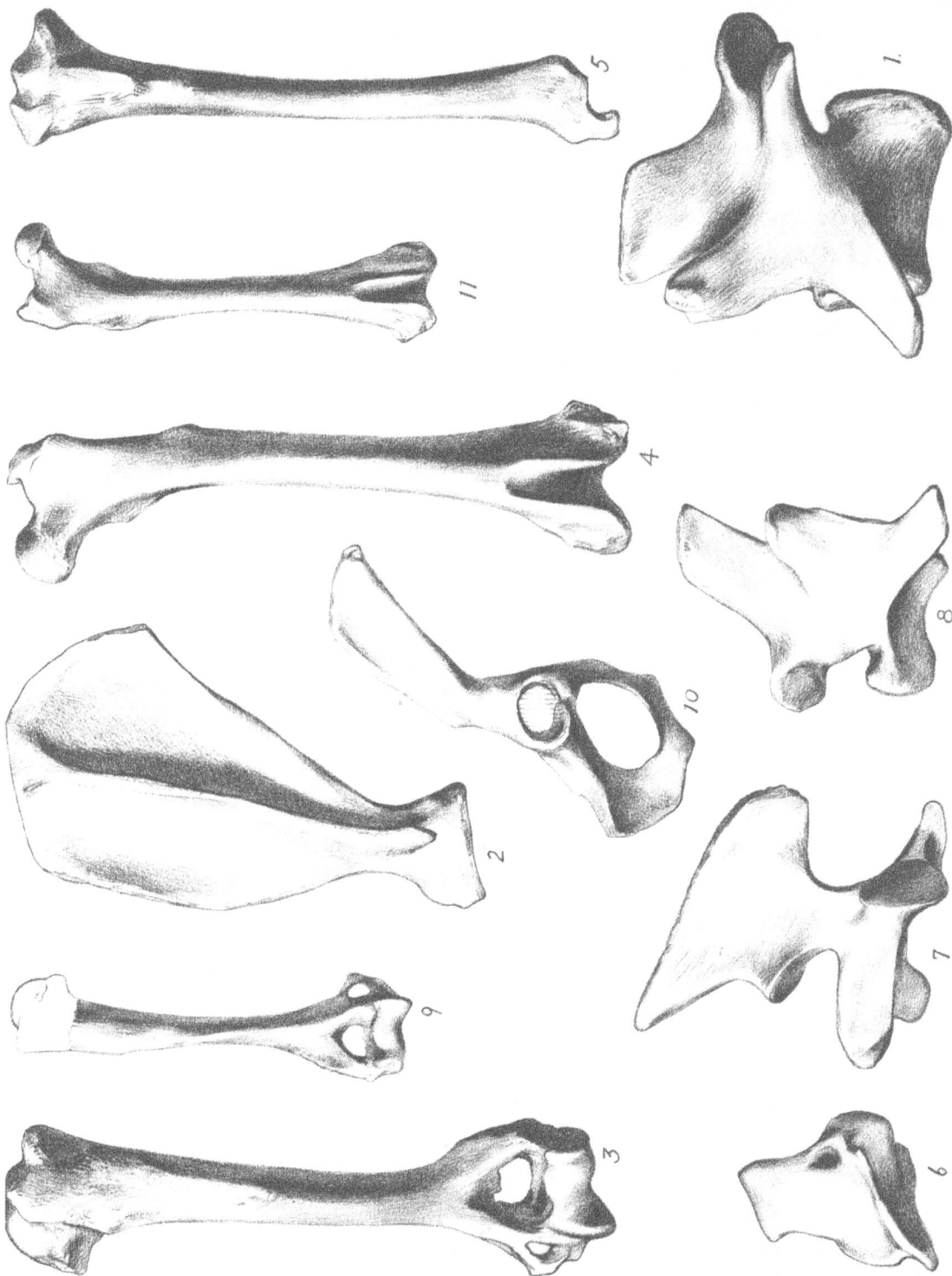

HYAENODON

PLATE X

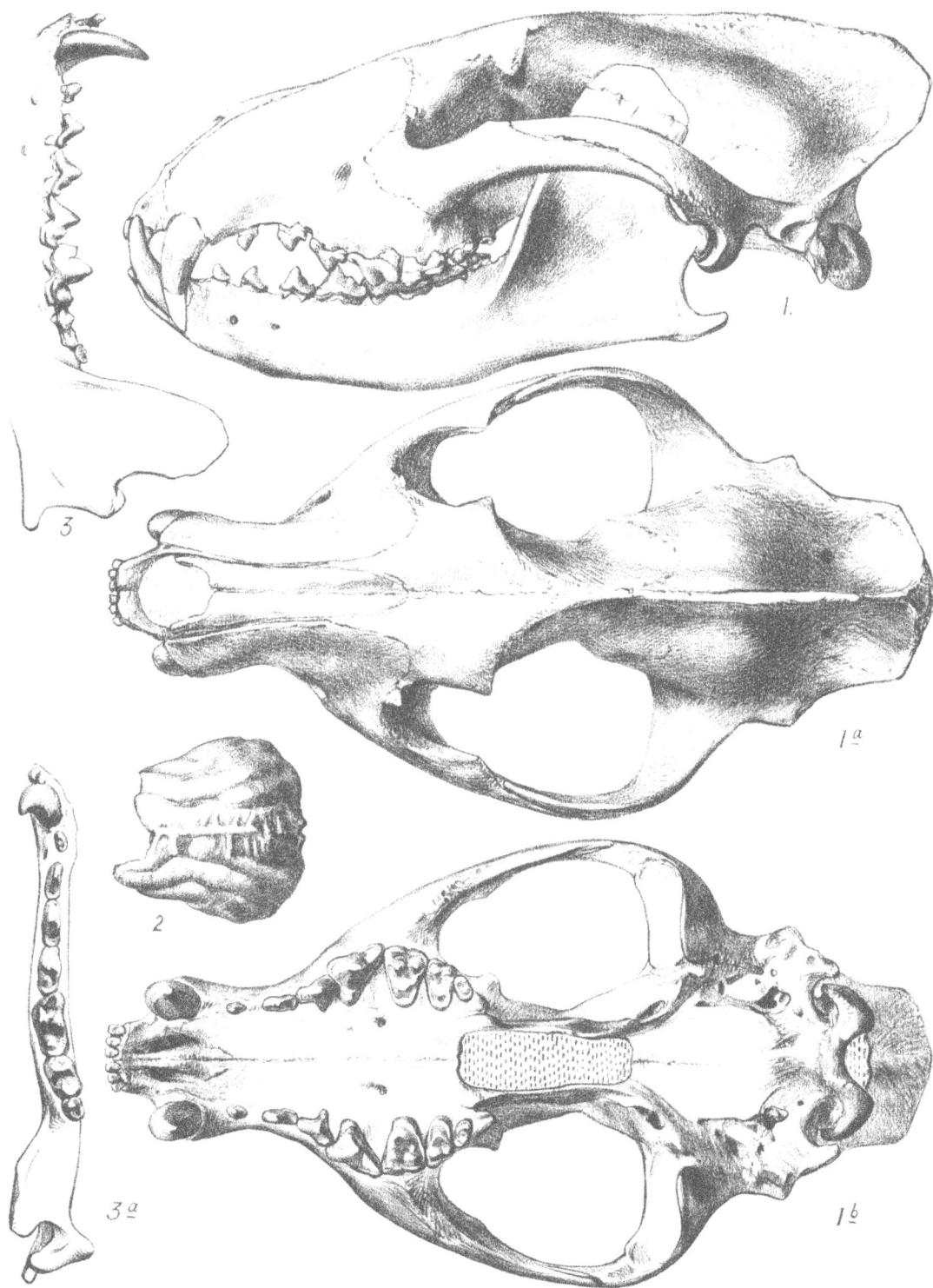

DAPHOENUS

PLATE XI

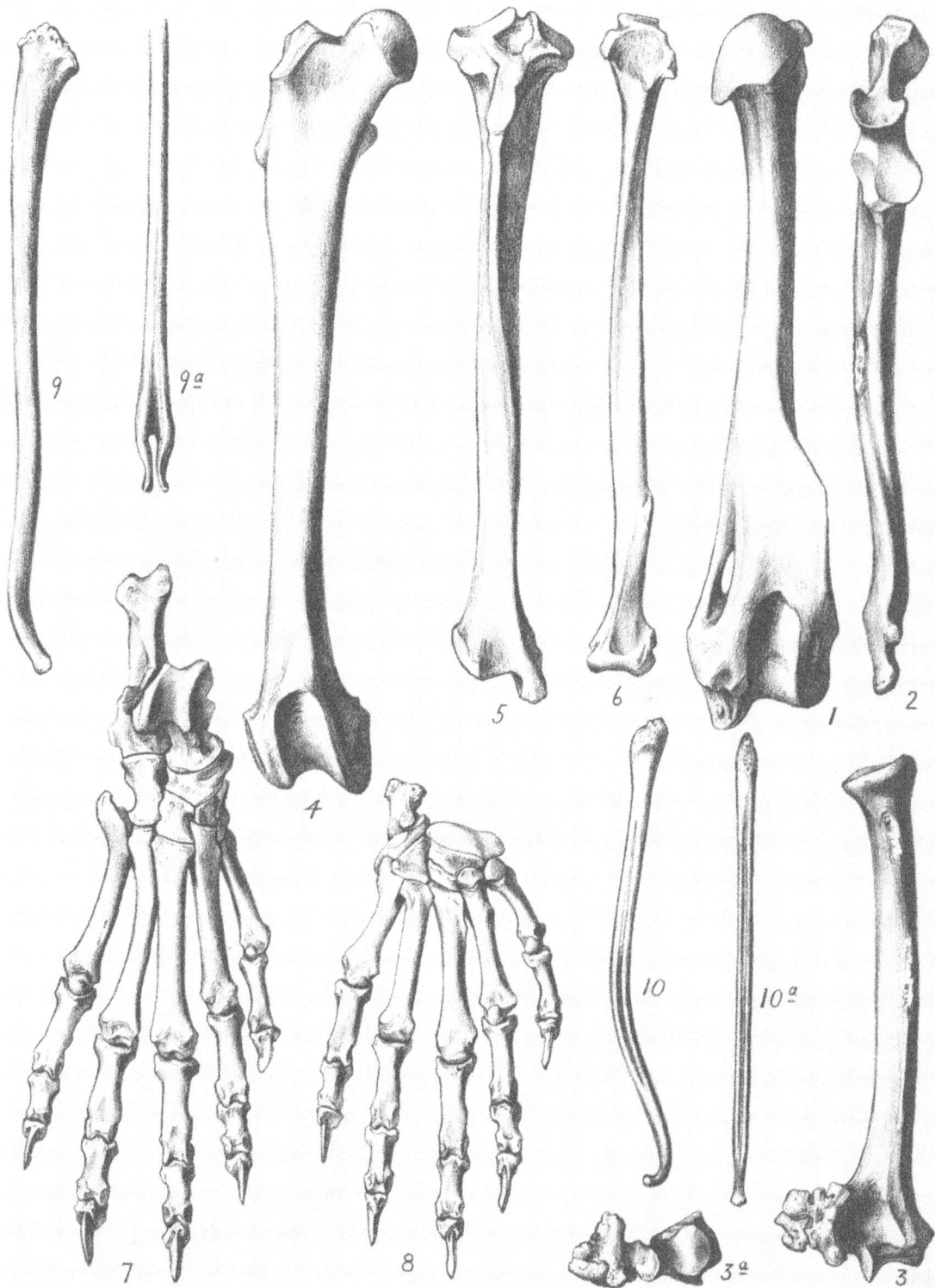

DAPHOENUS—PSEUDOCYNODICTIS

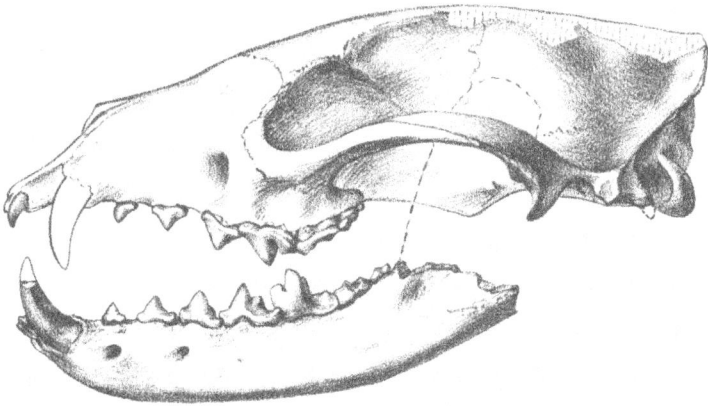

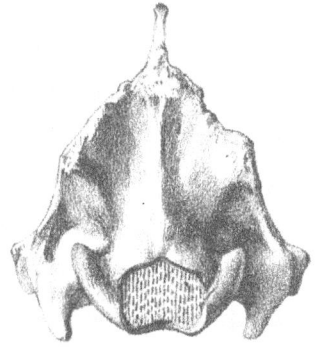

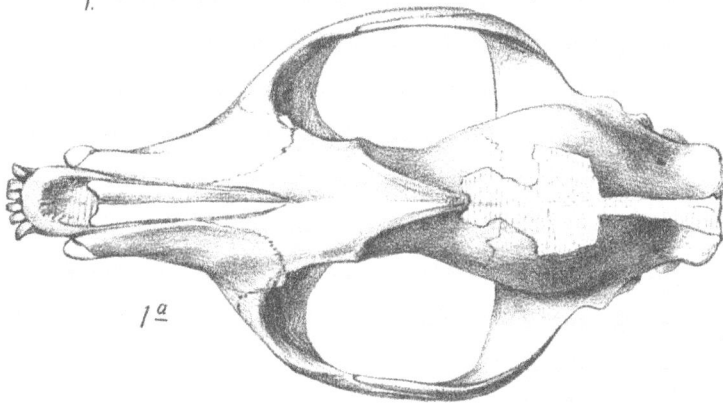

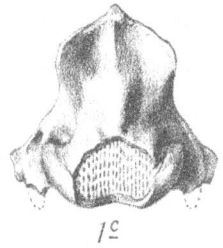

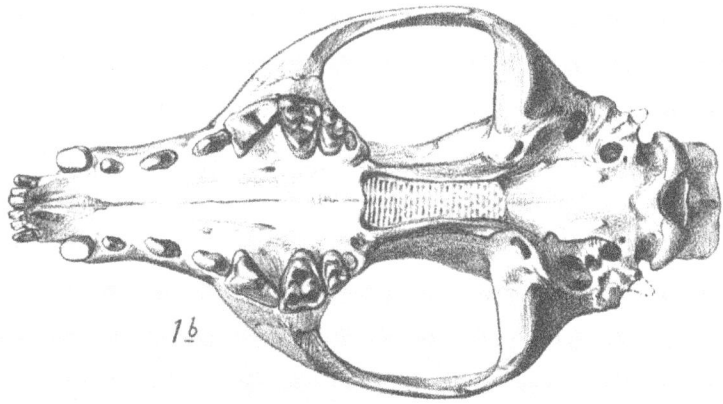

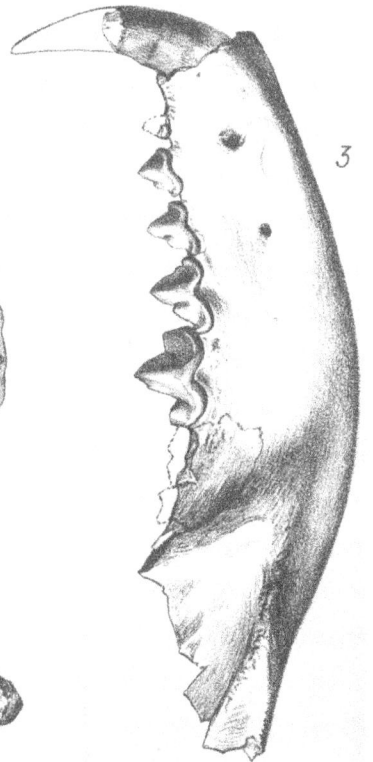

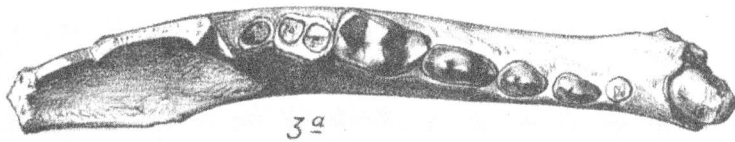

PLATE XII

1.

1ª

1ᵇ

2

1ᶜ

3

3ª

DAPHOENUS

PLATE XIII

1.

2.

3.

4.

1a.

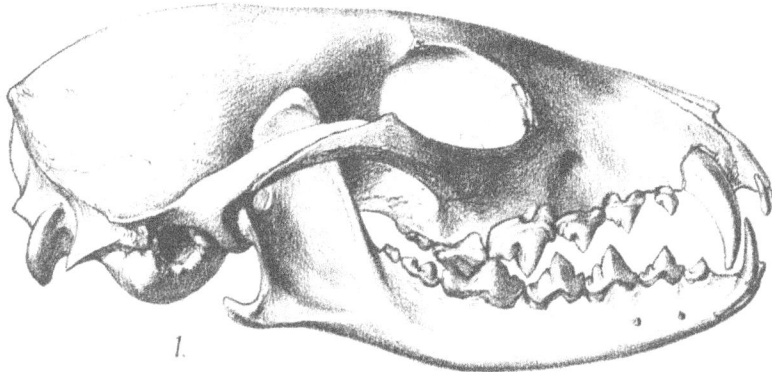

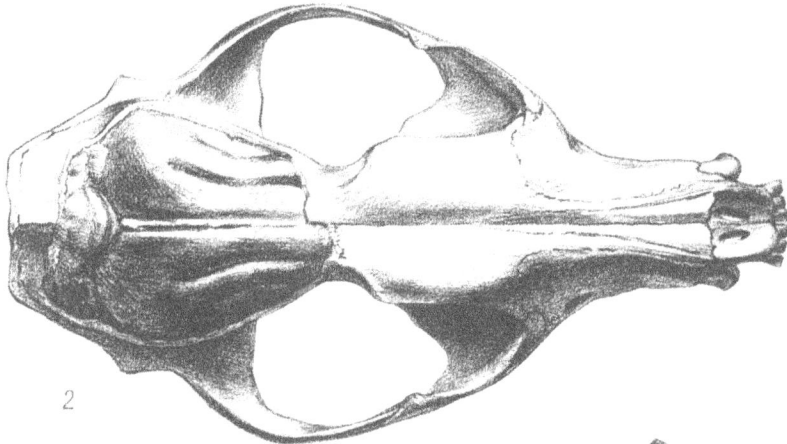

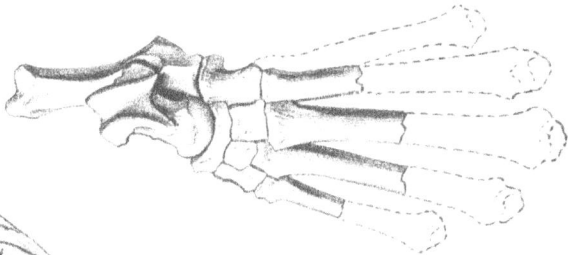

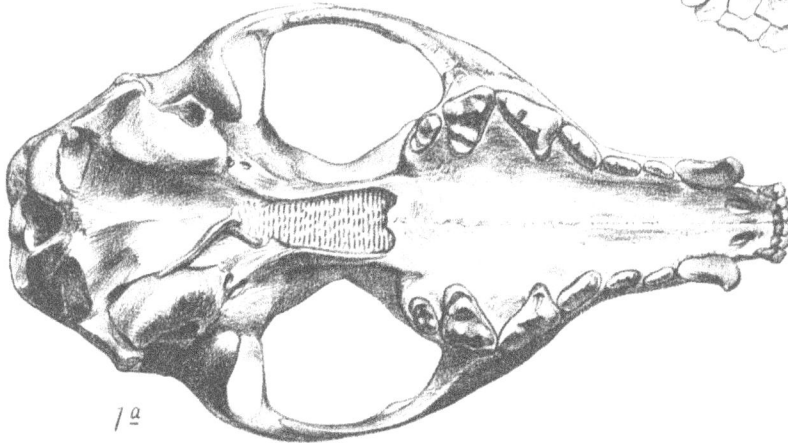

PSEUDOCYNODICTIS

PLATE XIV

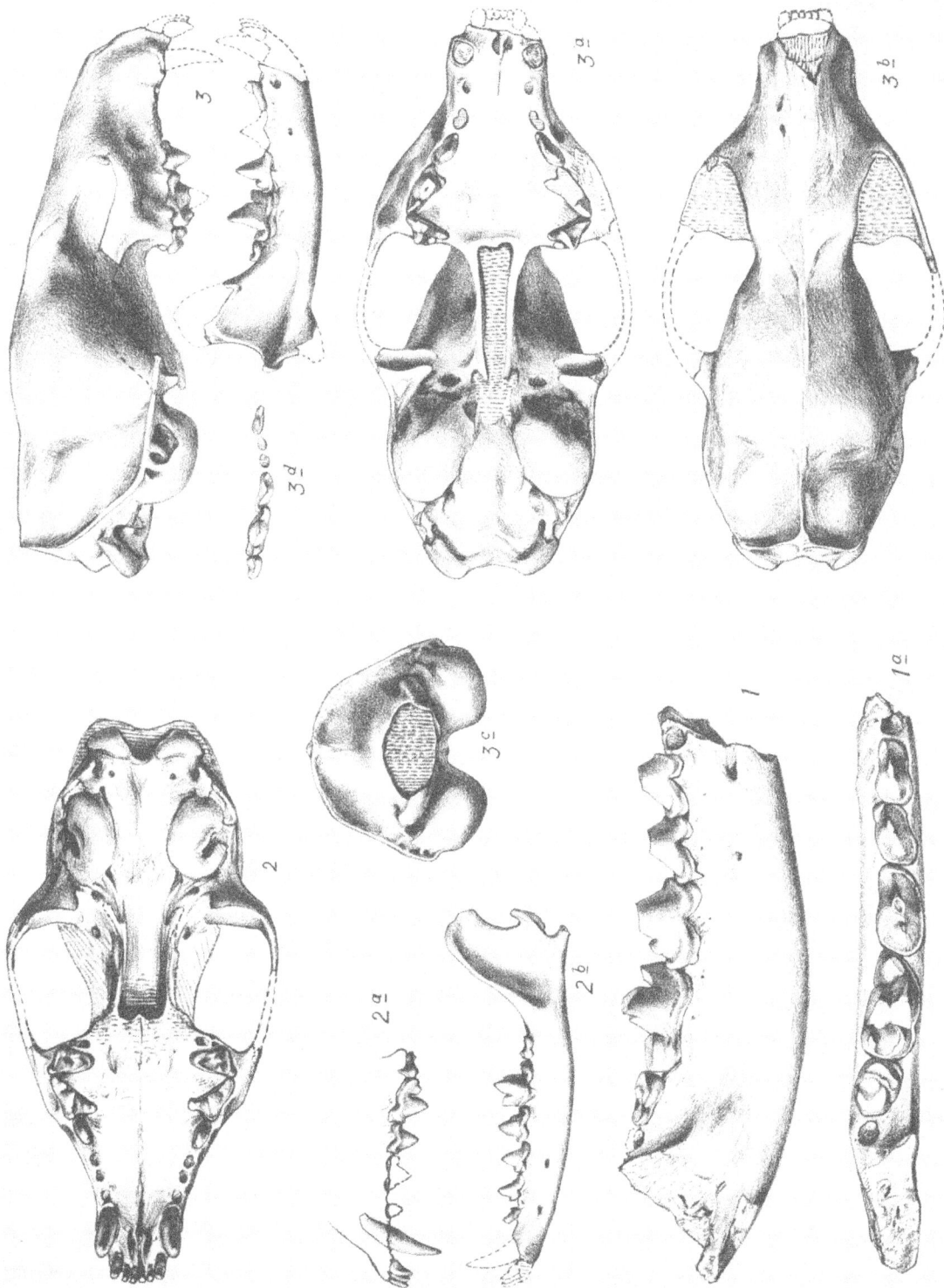

PARICTIS and MUSTELIDAE

PLATE XV

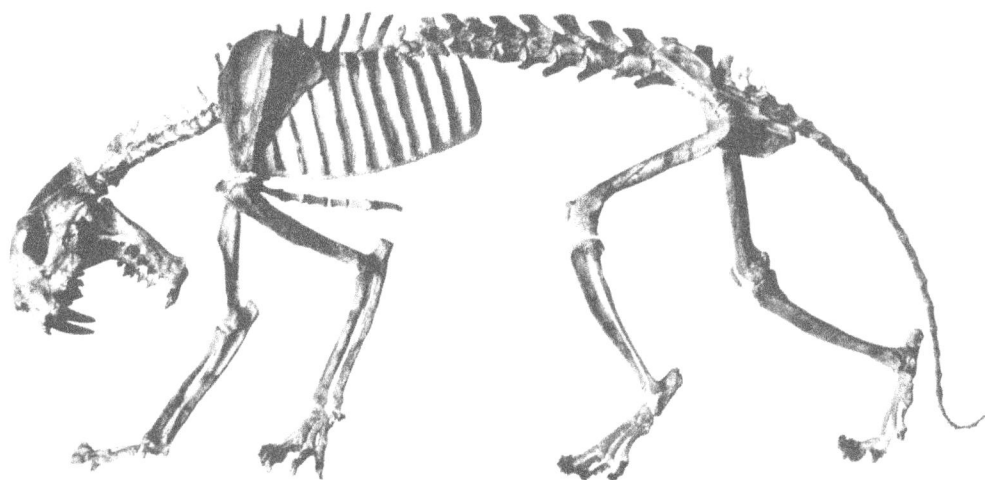

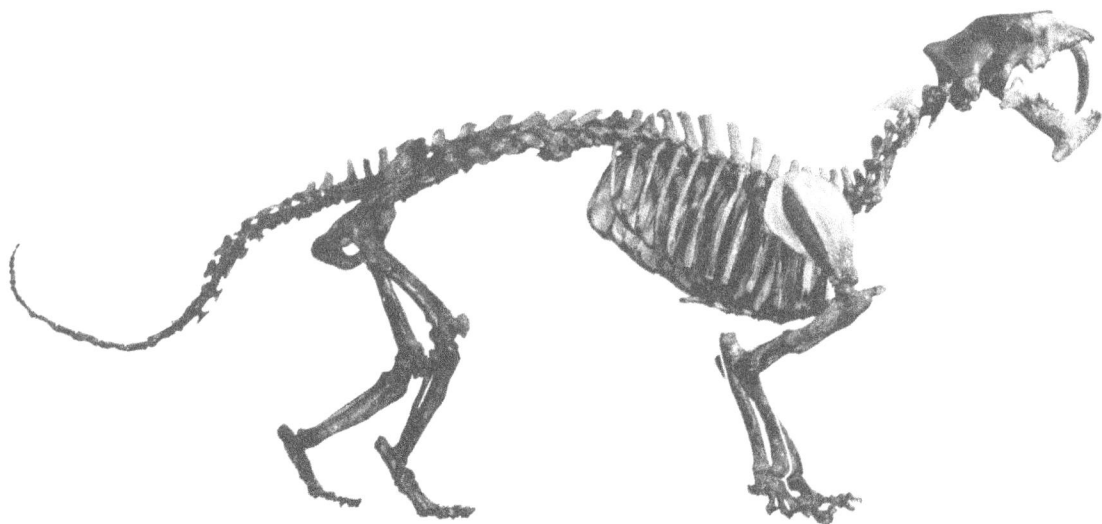

DEINICTIS—DREPANODON

PLATE XVI

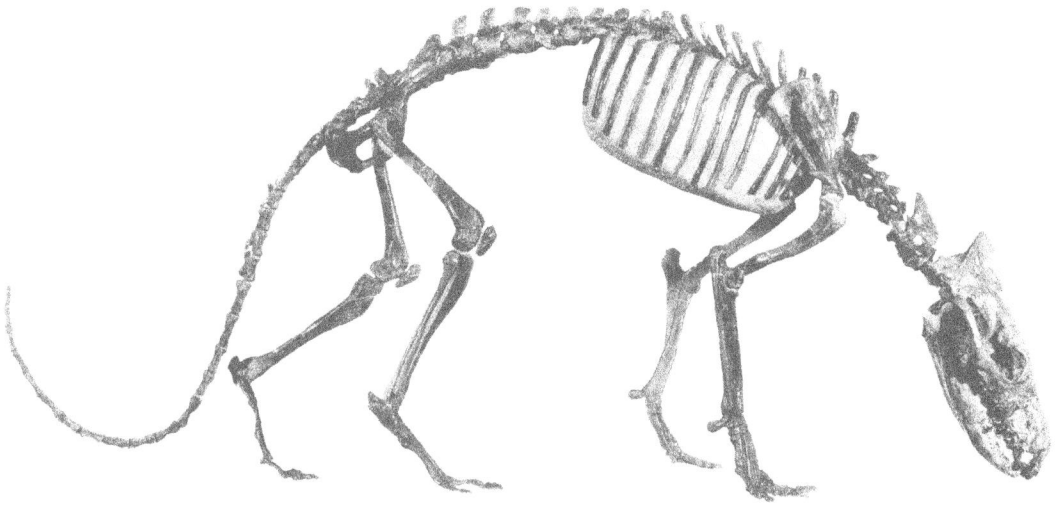

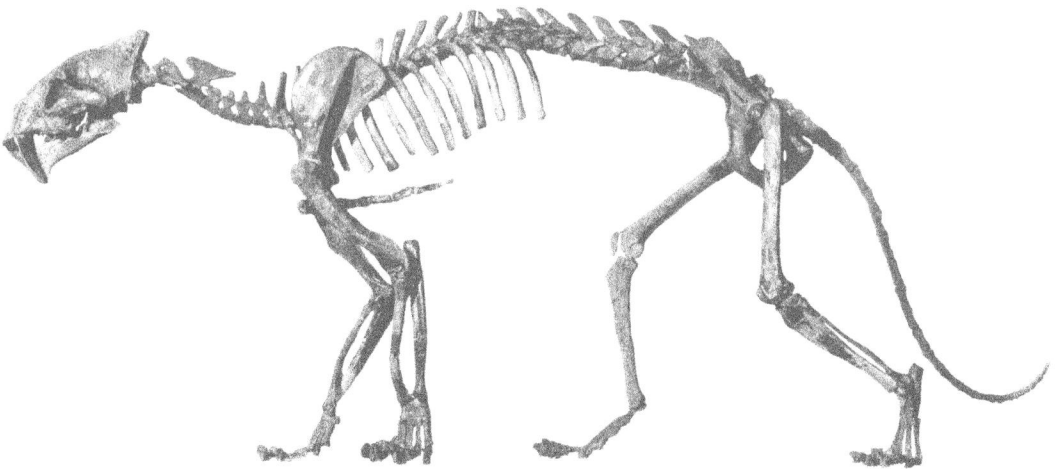

HYAENODON—DREPANODON

PLATE XVII

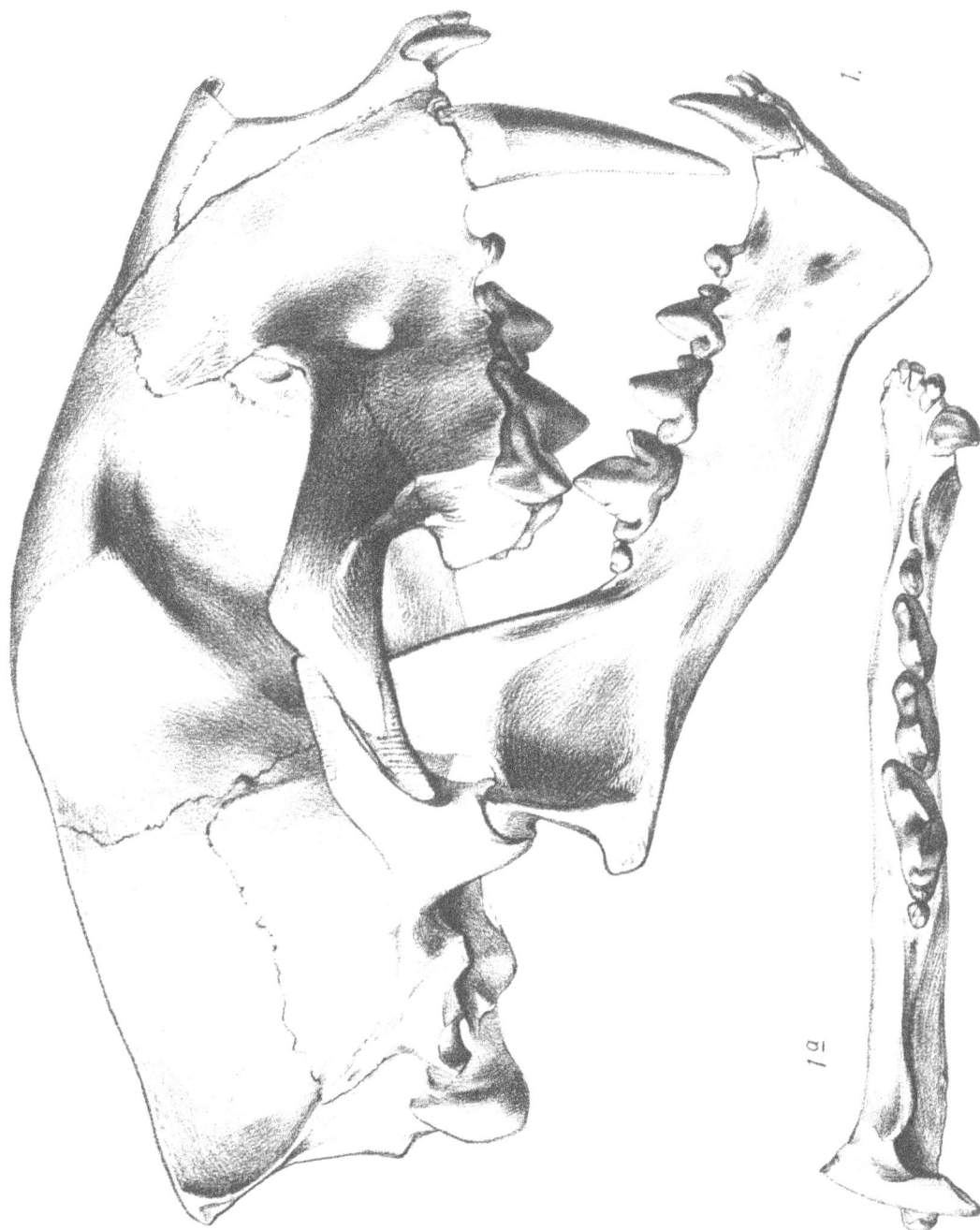

1.

1a

DEINICTIS

PLATE XVIII

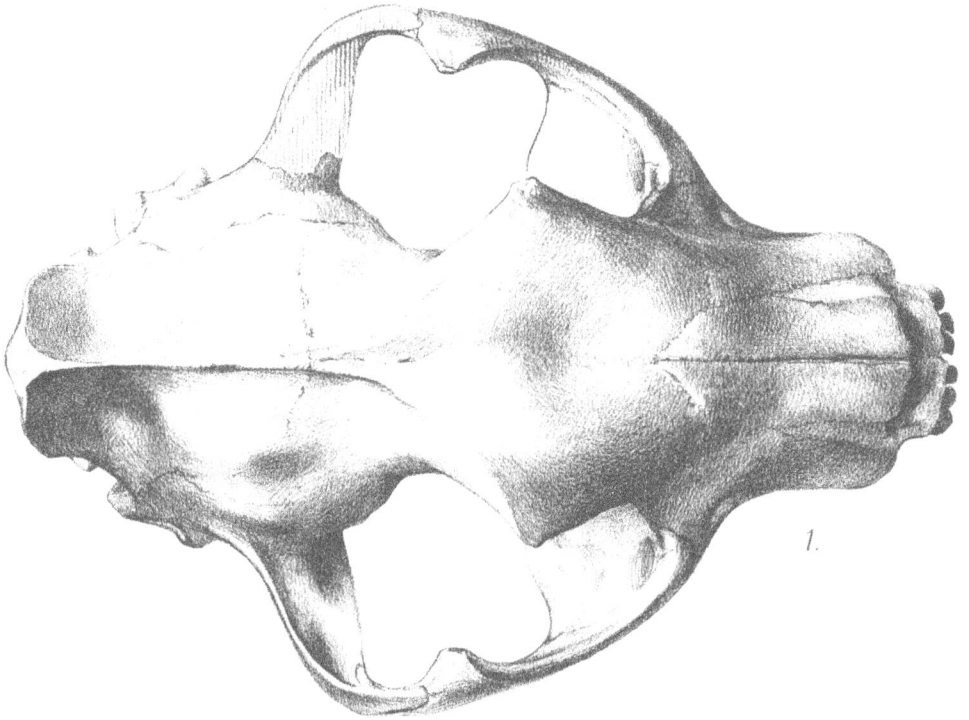

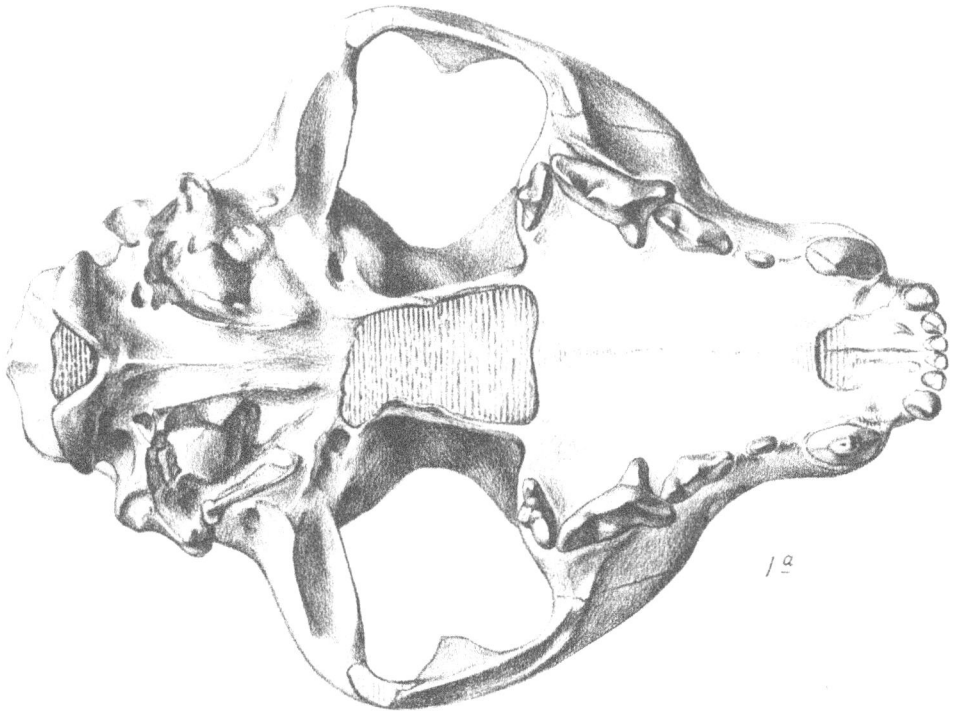

DEINICTIS

PLATE XIX

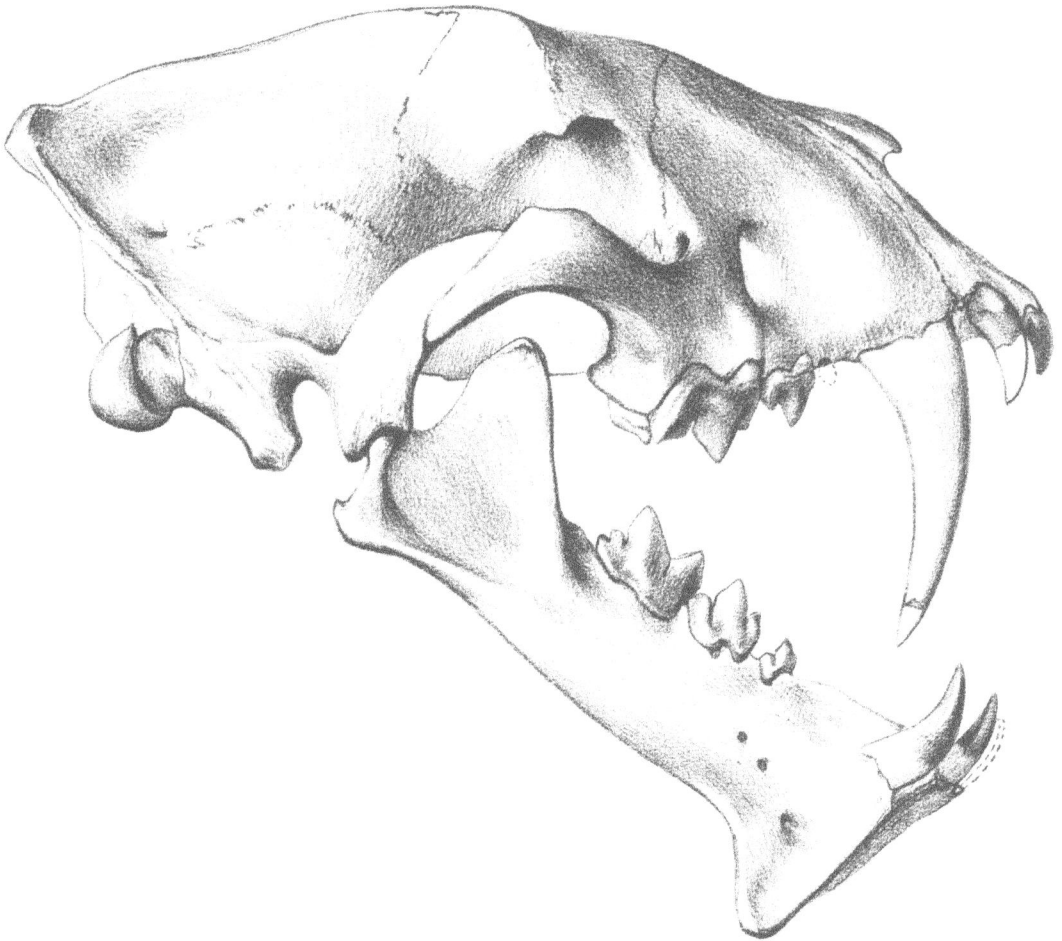

DREPANODON

PLATE XX

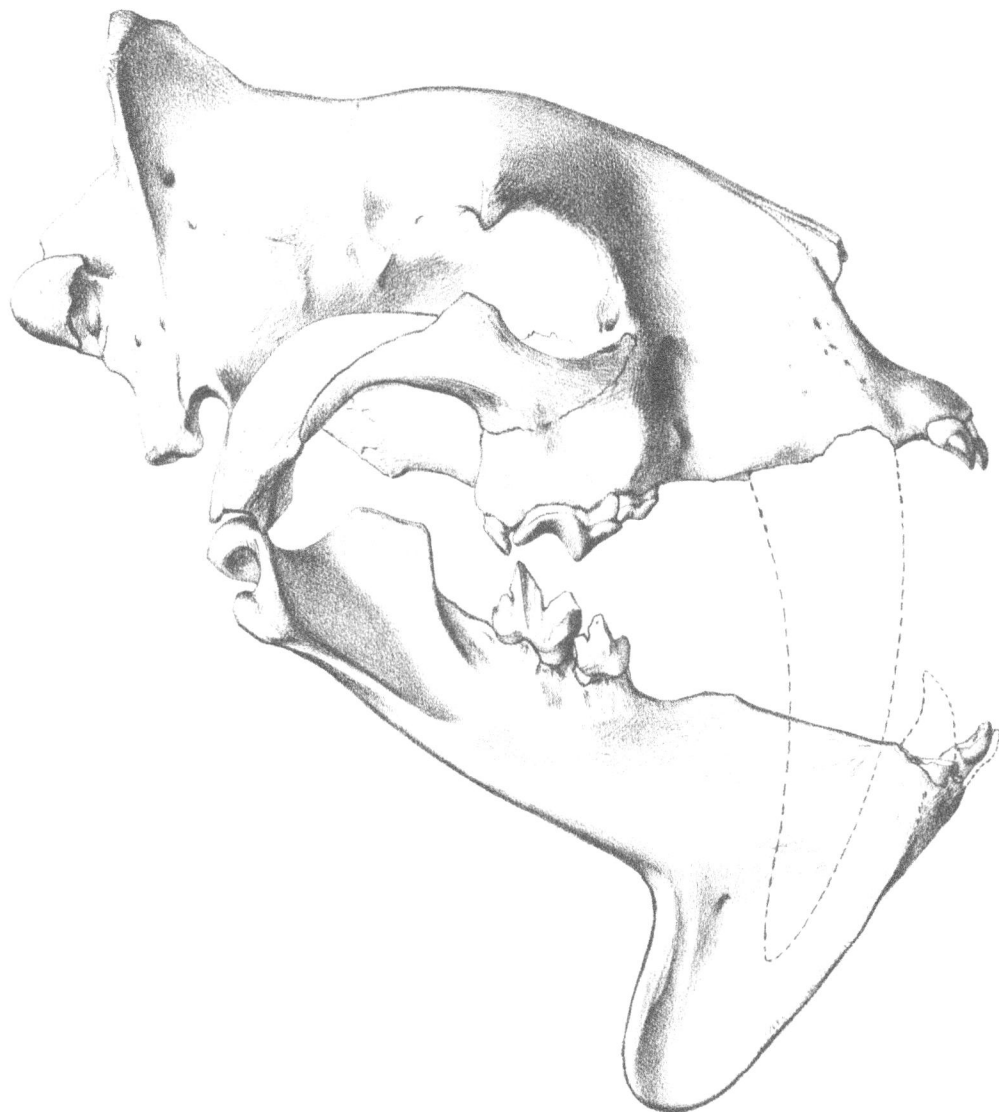

EUSMILUS

PLATE XXI

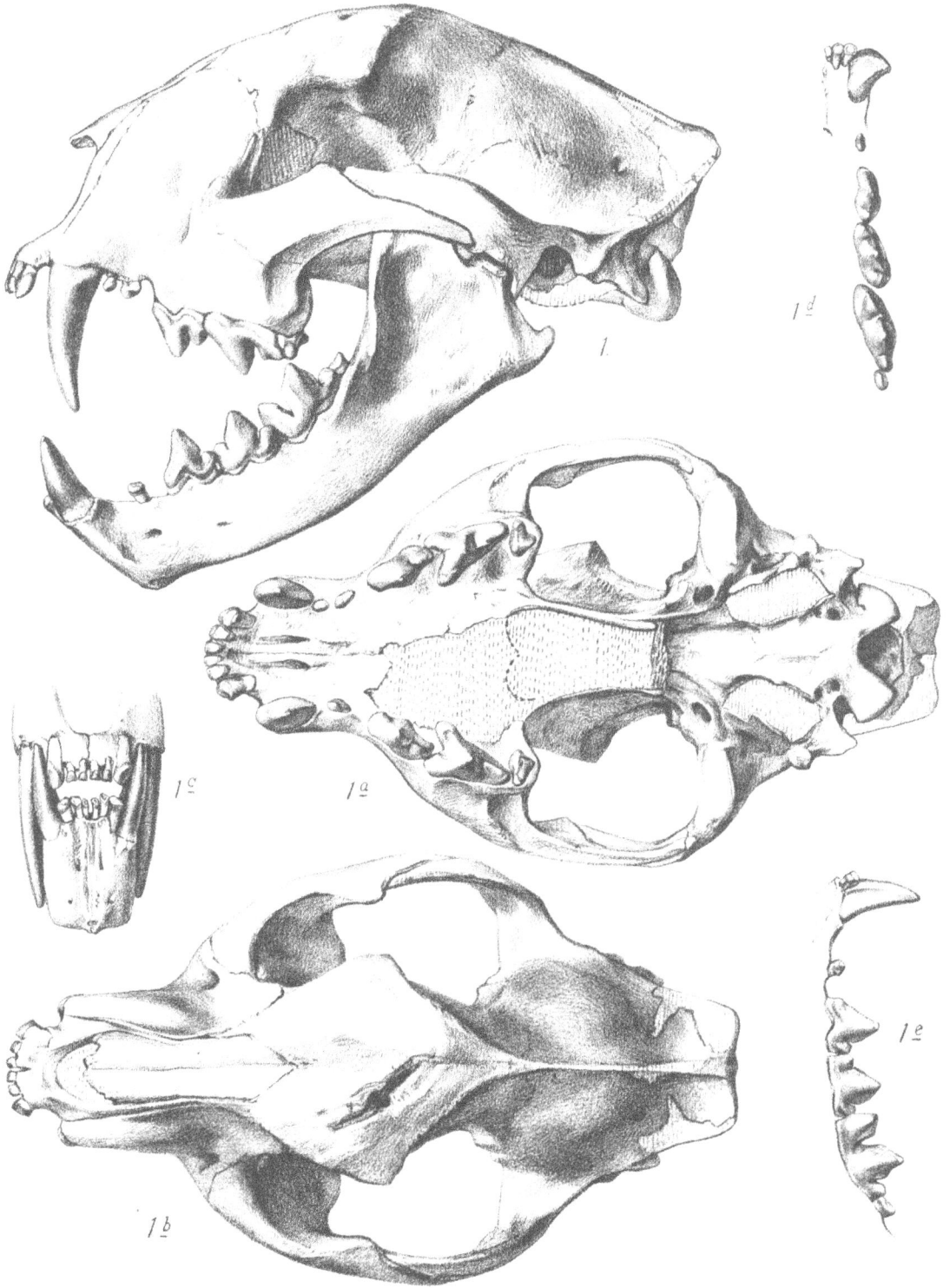

1.

1a

1b

1c

1d

1e

NIMRAVUS

PLATE XXII

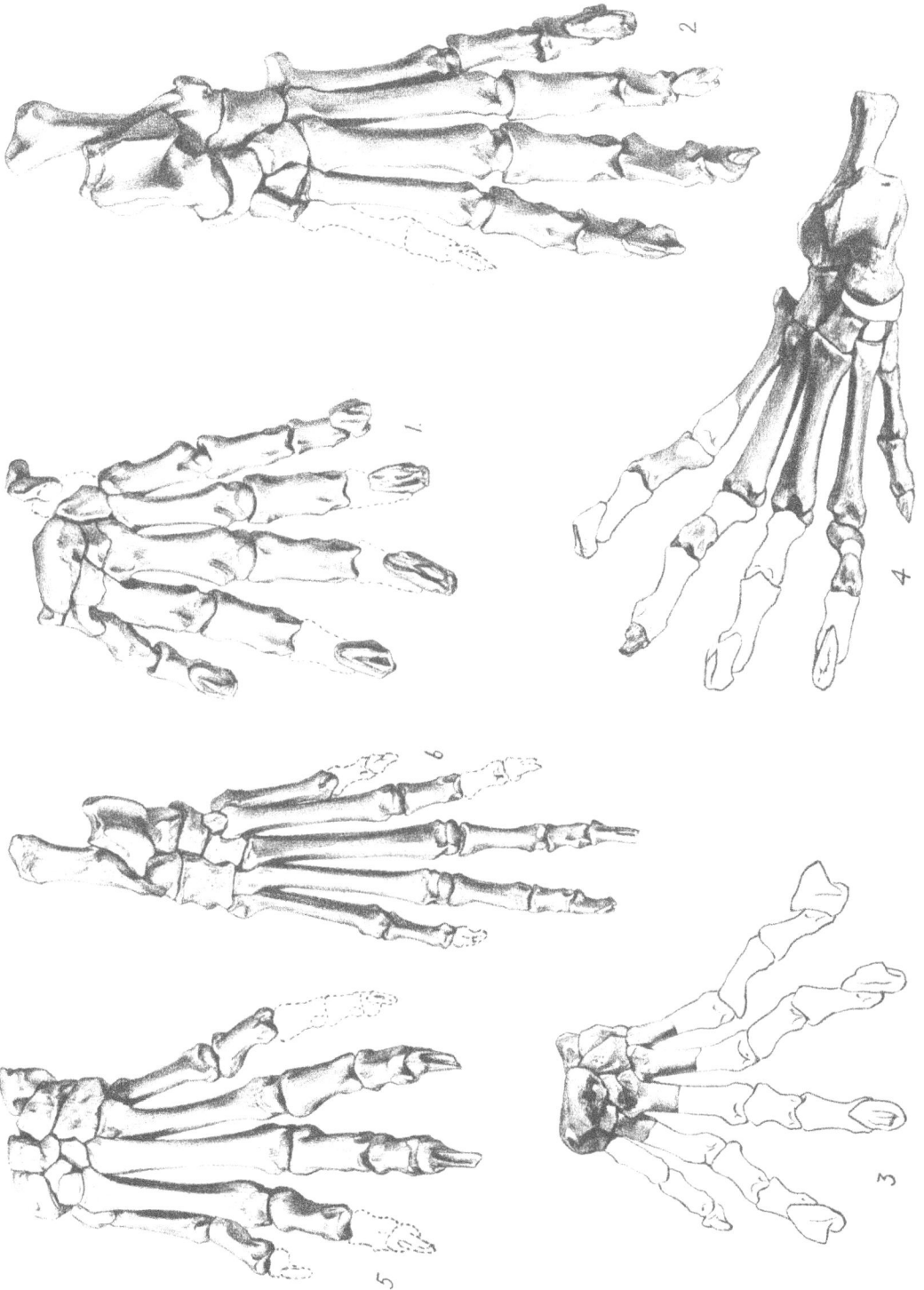

DREPANODON, DEINICTIS, HYAENODON

www.ingramcontent.com/pod-product-compliance
Lightning Source LLC
Chambersburg PA
CBHW081338190326
41458CB00018B/6036